Praise for *About T*

"Insightful, globe-spanning."   —James Gle..., ... of *Books*

"Fascinating. . . . [W]ith [David Rooney's] book in hand, and an eye on the world that sustains us, we might just save ourselves."

—Jonathon Keats, *Forbes*

"Engaging. . . . [T]he details of [Rooney's] innumerable examples are often very intriguing."   —Nick Romeo, *Washington Post*

"People say time is money, but David Rooney knows better. In this information-packed swoop through history and into the future, he exposes time's many identities along with the hidden agendas of clocks. Time is knowledge. Time is power. Time is faith. Time is destiny."

—Dava Sobel, author of *Longitude*

"Not merely an horologist's delight, but an ingenious meditation on the nature and symbolism of time-keeping itself. From the medieval hour-glass to the Doomsday Clock, from Jaipur to Jodrell Bank, from GMT to GPS, David Rooney ticks off time in a highly entertaining series of historical tales and parables which also give pause for thought and sometimes alarming reflections. I will never hear the pips, or ask 'what's the time?' in quite the same way again. A striking success."

—Richard Holmes, author of *The Age of Wonder*

"*About Time* is an utterly dazzling book, the best piece of history I have read for a long time. From sundials in ancient Rome to astronomical, water-driven, mechanical, and atomic timepieces used throughout history and across cultures, David Rooney has written the definitive book on these remarkable objects that give order to everyday life. It is a moving and beautifully written book that even takes us 5,000 years into the future with plutonium clocks ticking away beneath our feet. There will be many puns about this as a timely book; in fact, it is timeless."

—Jerry Brotton, author of *A History of the World in Twelve Maps*

"Splendidly thought-provoking. . . . [S]tartlingly original. . . . Rooney is immensely knowledgeable and passionate about his subject. His engaging style should make this book, which carries valuable warnings about the future of humanity, a popular-science classic."

—Patricia Fara, *Literary Review*

"The author knows his subject intimately. . . . [A] fascinating story about how clocks have not only kept the time for us but also defined the times we've lived in."

—Michael Taube, *Washington Examiner*

"A beguiling book that shifts enjoyably between the barely fathomable nature of time and historical anecdote." —Alan Connor, *Mail on Sunday*

"A fascinating and sometimes frightening story. . . . Rooney weaves a convincing tale."

—William Hartston, *Daily Express*

"It's to Rooney's credit that although he clearly knows a colossal amount about clocks, he wears his learning very lightly. . . . The details are fascinating."

—Dominic Sandbrook, *Sunday Times* (UK)

"David Rooney's passionate enthusiasm for everything clock-related leaps off every page. The vivid writing, engaging stories, and autobiographical details combine to offer a rich and generous picture of the history of clocks, from China and Japan to Central Europe, the Middle East, and outer space. In clear, pacey, and evocative prose, Rooney's volume takes in ancient wonders and modern marvels, leaving us at once enlightened and moved."

—Ludmilla Jordanova, author of *History in Practice*

"The measurement of time is a convenience, a jailor, a tyrannical device. David Rooney's delightful and discursive work anatomizes that tyranny. Page after page offers up instances of time's ubiquity and its mercurial power to get into the interstices of the everyday."

—Jonathan Meades

"Enthralling and important, *About Time* takes us deep into the past and far into the future. With David Rooney as personable guide, we peer inside clocks from Kyoto to Cape Town, discovering what they meant to the diverse people who made them, used them, whose lives were ruled by them. . . . This is a gripping and revealing account of time, and humanity's changing relationship with it."

—Seb Falk, author of *The Light Ages: The Surprising Story of Medieval Science*

About Time

About Time

A History of Civilization in Twelve Clocks

DAVID ROONEY

W. W. NORTON & COMPANY
Independent Publishers Since 1923

Copyright © 2021 by David Rooney
The credits on pp. 259–260 constitute an extension of this copyright page.
First American Edition 2021

First published in Great Britain by Viking, an imprint of Penguin Books

For information about permission to reproduce selections from this book, write to
Permissions, W. W. Norton & Company, Inc., 500 Fifth Avenue, New York, NY 10110

For information about special discounts for bulk purchases, please contact
W. W. Norton Special Sales at specialsales@wwnorton.com or 800-233-4830

Manufacturing by Lakeside Book Company
Book design by Chris Welch
Production manager: Lauren Abbate

Library of Congress Cataloging-in-Publication Data

Names: Rooney, David, 1974– author.
Title: About time : a history of civilization in twelve clocks / David Rooney.
Description: First American edition. | New York : W. W. Norton & Company, 2021. |
Includes bibliographical references and index.
Identifiers: LCCN 2021025193 | ISBN 9780393867930 (hardcover) | ISBN 9780393867947 (epub)
Subjects: LCSH: Clocks and watches—History. | Horology—History. |
Time—Social aspects—History. | Civilization.
Classification: LCC TS542 .R66 2021 | DDC 681.1/13—dc23
LC record available at https://lccn.loc.gov/2021025193

ISBN: 978-1-324-02195-7 pbk.

W. W. Norton & Company, Inc., 500 Fifth Avenue, New York, N.Y. 10110

www.wwnorton.com

W. W. Norton & Company Ltd., 15 Carlisle Street, London W1D 3BS

1 2 3 4 5 6 7 8 9 0

Contents

Illustrations

About Time

Introduction

Korean Air Lines Flight 007, 1983

It is the early hours of a crisp Alaskan morning. Korean Air Lines' Captain Chun Byung-in, First Officer Son Dong-hui and Flight Engineer Kim Eui-dong stride purposefully across the tarmac of Anchorage International Airport and climb into the cockpit of the Boeing 747 airliner that they are rostered to fly to Seoul's Gimpo International Airport.

Flight KAL 007 has stopped off at Anchorage on its journey from New York's John F. Kennedy International Airport for servicing, refueling and a changeover of the flight and cabin crew. The Alaskan airport, on the northwest tip of North America, is, at this time, a common staging post for flights between the USA and eastern Asia. Much airspace over the communist countries of Asia and Europe is closed to foreign traffic, meaning longer routings for flights seeking to find a way through safe international corridors. But Chun, the pilot of the flight, knows the passage from Anchorage to Seoul like the back of his hand, having flown it for half a decade.

The first leg of flight KAL 007 has been uneventful for the 269 people on board, and weather conditions for the second leg are predicted to be good, with lower than average headwinds meaning the flight duration will be slightly reduced. In order to arrive at Seoul on time, the departure from Anchorage is therefore set back by half an hour. The final checks are completed and nothing seems out of the ordinary. A route is punched into the navigation computer that will take the aircraft safely around the outer edges of prohibited airspace, and the airport's radar systems record flight KAL 007 in the air at 4 a.m., Alaska time. It has all the makings of an unremarkable flight.

The hours pass. Conversation among the flight crew is jovial and relaxed. At certain points during the flight, they contact ground controllers to report their position and weather, and to confirm plans. Breakfast is served to the passengers, just as normal.

But there is a problem with the aircraft's autopilot. What Chun, Son and Kim have not realized is that it has not been set up correctly, and throughout the course of the flight from Alaska they have strayed increasingly to the north of their intended route. It is the worst possible mistake they could have made. With no way to double-check their position, they have relied on their navigational equipment to direct the aircraft along the required route, but it has taken them directly into prohibited airspace over the Kamchatka Peninsula and the island of Sakhalin.

Five hours after the Boeing jet leaves Alaska, and unknown to the Korean flight crew, a Sukhoi Su-15 supersonic jet, piloted by Lieutenant Colonel Gennadi Osipovich, is scrambled to intercept the airliner. Osipovich's military commanders have recently spotted a US spy plane operating in the area, monitoring a missile test being carried out. This is a well-known Boeing RC-135 four-engine reconnaissance jet, similar in many ways to the Boeing 747 passenger jet but without the distinctive hump above the cockpit. Osipovich and his bosses are convinced the Korean Air Lines aircraft is another US spy plane.

Twenty minutes later, having reached the airliner with its oblivious crew and passengers, Osipovich fires a burst of warning shots from his cannon across the Boeing's nose, but the shells cannot be seen by the Korean crew, who carry on chatting, unaware of the danger that is fast closing in on them. Six minutes after that, Osipovich launches two air-to-air missiles at the Korean airliner. One misses, but the other explodes at the Boeing's tail, severing hydraulic control lines and inflicting significant structural damage. Shrapnel from the blast penetrates the airliner's fuselage, causing the cabin to decompress. Though it has been mortally wounded, flight KAL 007 continues to fly onward as the crew struggles to regain control. Automated announcements on the public-address system begin to sound throughout the aircraft, thirty seconds after the missile strikes. "Attention. Emergency descent. Put out your cigarette. This is an emergency descent." Oxygen masks drop from the ceilings in the cabin and cockpit and the PA system begins to shout, "Put the mask over your nose and mouth and adjust the head band. Attention. Emergency descent."[1]

The airliner continues to hurtle through the skies above the Sea of Japan. The passengers who remain conscious, while they do not know what hit them, or why, are in no doubt about the grave danger they face if the aircraft cannot be brought into an emergency landing. The crew continue to wrestle valiantly with the controls as they become less and less responsive. The airplane bucks and rolls as it is buffeted by winds and weather, having lost the aerodynamics needed for safe flight. Twelve minutes after the missile is fired, what limited control the pilots have of the jet is lost and, having plummeted down in a deadly spiral, flight KAL 007 slams into the ocean. The terror is finally over. It is the morning of September 1, 1983, and there are no survivors.

Overhead, a fleet of seven experimental US military satellites called Navstars is orbiting. Each satellite is the size of a family automobile and weighs just short of a ton. They are powered by a combination of solar cells and hydrazine rocket fuel, and the fleet has been launched, one by one, every few months since 1978. Between them, these satellites are carrying twenty-five high-precision clocks, built in California, as part of a navigational experiment called the Global Positioning System.

These clocks could have saved everyone on board flight KAL 007.

––––•·–––

FOUR DAYS AFTER the Korean airliner was shot down by a Soviet missile, the US president, Ronald Reagan, made an emotional television
address in which he described the tragedy as a "massacre," a "crime
against humanity" and an "act of barbarism" by the Soviet authorities,
vowing to take steps to ensure it never happened again.[2]

The experimental satellites flying above the aircraft as it plummeted
down to Earth were the first in a constellation we know today as GPS,
then being developed by the US military. Each GPS satellite carried three
or four miniature atomic clocks, which beamed precise time signals to
Earth, where people carrying GPS receivers could find their position to
within tens of yards. Today, the GPS system involves around thirty-two
satellites that are active at any one time, and the latest ones carry clocks
far more reliable and accurate than the first ones, made in the mid-1970s.

These space clocks have now become an invisible part of our everyday
lives, providing not just precise locations but synchronizing all modern
infrastructure, from telecommunications to power supply. In September
1983, the experimental GPS satellites were used only by the military. But
the downing of KAL 007, with the loss of 269 innocent lives, changed
that. Eleven days after his television address, Reagan announced through
his press secretary that civilian aircraft would be allowed to use GPS
when it became operational. If the experimental time signals had been
available to the Korean pilots, they might have alerted them to their navigational error and averted the tragedy of September 1, 1983.

These rather austere 1970s clocks, manufactured in a joint venture
between the American firm Rockwell and the German clockmaker Efratom, housed in rugged aluminum boxes and hardened against the batter-
ing they would receive when launched into space, may not match up to
our mental image of fine, important clocks. They are not conventionally
beautiful, and there are few collectors willing to give room in their houses
for them. Yet they have changed the world, not just technically, but politi-

cally and culturally. They are clocks placed above our heads by a military superpower. The service they provide is not—was never—benign. Should we not, therefore, consider them more critically than we do?

The original 1970s clocks are still with us. Superseded by more recent technology, the twenty-five clocks on those first seven GPS satellites, which were orbiting Earth as flight KAL 007 plunged into the Sea of Japan, are nevertheless still in orbit around the Earth. These are real clocks, made by clockmakers in factories like Rockwell's and Efratom's in California, but now shut down, drifting eternally and silently over our heads on long-retired satellites. The night sky is a museum of old clocks, if we could see that far.

———

SINCE THE EARLIEST civilizations, people of all cultures have made and used clocks. From the city sundials of ancient Rome to the medieval water clocks of imperial China, and from hourglasses fomenting a quiet revolution in the Middle Ages to Enlightenment observatories in India, a history of clocks is a history of civilization. This book, then, is for anybody interested in the history of the world, in politics, and in how the story of timing is the story of us. It will explore twelve case studies— twelve real clocks from our past—to show how, for thousands of years, time has been harnessed, politicized and weaponized. With clocks, the elites wield power, make money, govern citizens and control lives. And sometimes, also with clocks, people fight back. None of this is abstract. These are real clocks with recoverable histories that bring pivotal and sometimes violent moments from the past vividly to life.

———

MY FASCINATION WITH clocks and their history started at a young age. In 1982, when I was eight years old, my parents decided to set up a clockmaking and restoration business. My mother had been a researcher for

Tyne Tees Television in the mid-1960s before becoming a teacher. My father had been an engineering draftsman at the Hebburn-based firm of Baker Perkins, before he, too, started teaching. But they had always yearned to run their own business and, in the early 1980s, took the plunge. They worked from our family home, a terraced house in South Shields, on England's cold North Sea coast as it meets the River Tyne. We lived, as it happens, close to the old Harton Pit, a former coal mine that in 1854 was the scene of pioneering experiments by the nation's top time scientists using pendulum clocks to study the density of the Earth. Clocks had been the talk of the town in nineteenth-century South Shields.

Our dining room was converted to a horological workshop and library. A spare bedroom became the office. The kitchen table, where we ate all our meals, was where as a child I picked up the language of clocks and watches, hearing discussions about the arcane technology of horology— fusees, escapements, oscillators—as well as the challenges of working with these complex machines and of running a business. I heard about my parents' dealings with horology's noted scholars and collectors, and I often accompanied them to set up clocks in country houses and museums across Scotland and the north of England.

I suppose I absorbed a hybrid combination of my father's technical appreciation of clocks and my mother's experience in research for television documentaries. Both my parents understood how important it was to tell their clients the *stories* of the clocks they worked on. It was never just a matter of fixing them up. Each clock had a life story and was part of history, however modest; it was my parents' job to find this story and share it.

Having lived through a decade of clockmaking and its history as a child, I left for university, where I studied physics, and later the history of science and technology while I was working as a technology curator at the Science Museum in London. At the Royal Observatory in Greenwich, where I became curator of timekeeping in the mid-2000s, I was given unfettered access to one of the world's most remarkable collections of precision clocks and watches. Three days a week, I wound the celebrated marine timekeepers made by John "Longitude" Harrison and helped look after the observatory's time ball and its pioneering Victorian electrical time network. And I volunteered every month at Belmont, a country house in Kent which holds one of the world's finest private collections of clocks and watches.

I was hooked. Later, back at the Science Museum, I looked after its own horology collection, among others, and started collaborating with the Worshipful Company of Clockmakers over its museum, the oldest of its type in the world, which moved to South Kensington in 2015. I also benefited over the years from the company, wisdom and patience of countless clock, watch and time specialists who generously shared their knowledge and passions with me, and still do. Throughout all this, my interest in these remarkable devices has only grown.

What fascinates me most is what clocks *mean*, a question which is answered by looking at why people have made them. The more I have learned, the more it has become obvious that the technical history of horology is only the start of the story. It is human motivation and how the world works that really interests me, which is why this is a story centered on power, control, money, morality and belief.

———•———

IT SHOULD BE apparent by now that this is not a conventional history of clocks and watches; nor does it deal with the more abstract concept of time itself—what philosophers and scientists think time *is*. There are many books which do this far better than I ever could, and I will therefore leave it to the experts. Nor is this a broad, sweeping account of the history of civilizations, like the fine works of the French historian Fernand Braudel or any number of other great scholars. Instead, it is a personal, idiosyncratic and above all partial account. It looks at how we can understand our history better if we examine artifacts that, for one reason or another, shed light on aspects of civilizations that matter to us. These aspects include the ways we are governed; the beliefs we hold; and the ways we tell stories. We will use the history of clocks to look at capitalism, the exchange of knowledge, the building of empires and the radical changes to our lives brought by industrialization. We will consider morality—right and wrong—as well as identity—who we are—all mediated by clocks. And we will look unflinchingly at life, death, war and peace. People use clocks to kill us, but clocks might save us, too, if we would only think about the power they wield.

The word "clock" will be used very loosely throughout *About Time*. It derives from European words meaning "bell" such as *cloche*, *Glocke* and *klocka*. Today, we tend to use it to mean fixed devices, either electronic or involving intermeshing geared wheels, that keep time and show it to us. I use it to mean far more. Throughout what follows, any human-made device with the purpose of tracking the passage of time is included in my definition of a clock. This includes sundials, hourglasses (sometimes called sandglasses or sand clocks), water clocks, time-finding telescopes, time signals, pocket watches, wristwatches—whatever.

———

ENOUGH INTRODUCTIONS. LET us begin our odyssey. And to do so we will travel back in time to ancient Rome, over 2,000 years ago, and look at a sundial, fixed to a column at the heart of the Roman Forum. The sundial has long since been lost, but, as you will see, its story is one that could have been written yesterday, such was the modernity of the concerns it raised at the time. Because the people of Rome were *not* happy with the way this sundial controlled their lives.

Order

Sundial at the Forum, Rome, 263 BCE

Everyone in Rome remembered the day the sundial came to town. Manius Valerius Maximus, the returning war hero, had stood proudly and imperiously on the elevated rostrum at the heart of the Roman Forum. In front of him were huge, cheering crowds, eager to celebrate their elected consul who had commanded Rome's military forces to a decisive victory on the island of Sicily. It was Valerius who had captured the city of Catania for the Roman Republic, and it was he who brokered a treaty at Syracuse, the most important strategic alliance in Roman history. The year was 263 BCE, and the taking of Catania an early success in the First Punic War, between the rival states of Carthage and Rome. War booty, plundered from the island, brought the victory tangibly back to the people. Often, that meant the prows of captured enemy ships, hacked off and mounted on columns in public centers like the Forum. But it was not all about military trophies and plundered treasure. One of the objects that Valerius had looted from Catania was modest to look at; even mundane. But it came to change the lives of ordinary Romans—and our own—forever.

Artist's impression of speakers in Rome's
Forum, published in 1851

Pointing to a spot by the rostrum on which he stood, Valerius revealed
the sundial he had brought back from Sicily and mounted on a column
that bore his name. It took the form of a large block of marble in which
a hemispherical cavity had carefully been chiseled out. At the top of the
cavity was a bronze pointer, or gnomon, and lines carved into the marble
acted as the time-telling scales onto which the gnomon's shadow fell. It
told the time and calendar of Sicily, slightly different from that of Rome,
but it did not really matter. It showed that Rome was on top, and the
crowd had gone wild.

Everybody knew that triumphal columns in public spaces like the Forum were symbols of great military power, which meant that Valerius' public sundial of 263 BCE, which was Rome's first, was not simply an ornament. As war booty from the sacking of Catania, displayed on its column in the very spot where Rome's most famous public speeches were made, Valerius' sundial stood proudly for the military might of the Republic. But this column was destined for higher fame. As the crowd dispersed from the Forum that day, few realized the true significance of what they had just witnessed. It had seemed as if they were celebrating a decisive victory against the Carthaginians by cheering the plundered sundial set up in Valerius' name. But they soon learned otherwise.

The sundial from Catania was joined by dozens more across Rome, each designed to regulate and control the myriad daily activities of Rome's citizens—who quickly became uneasy at the intrusion of this new timekeeping technology.

Things eventually got so bad that sundials became a target for the city's playwrights and critics, who poured scorn on the new devices. Writing a few years after the Forum sundial had first been installed, one exasperated playwright made a character exclaim:

> The gods damn that man who first discovered the hours, and—
> yes—who first set up a sundial here, who's smashed the day into
> bits for poor me! You know, when I was a boy, my stomach was the
> only sundial, by far the best and truest compared to all of these. It
> used to warn me to eat, wherever—except when there was nothing.
> But now what there is, isn't eaten unless the sun says so. In fact
> town's so stuffed with sundials that most people crawl along, shriv-
> elled up with hunger.[1]

A later writer described sundials like the one mounted at the Forum as "hateful" and called for the columns on which they were fixed to be torn down with crowbars.[2]

But it was too late. Public sundials began to appear across the Republic. Valerius' own triumphal sundial survived the public outrage for exactly ninety-nine years, only to be replaced in 164 BCE by one that was even more accurate. Five years after that, the hated sundials of Rome were joined by a new public timekeeper at the Forum, a water clock, which kept the time through the night as well as the day. Now the clock ruled the sleeping hours of Romans as well as their waking ones.

We should think of the sundial in the Roman Forum as the city's first clock tower. Mounted up high, looking over the people, and standing for Rome's ruling classes themselves, it changed everything. From the moment Valerius revealed his sundial at the Forum, Romans were forced to live their lives by the clock. And this new temporal order was sweeping civilizations across the world.

———

THE TOWER OF the Winds, in the Greek city of Athens some 650 miles from Rome, is one of the best-preserved buildings from the ancient world. This octagonal marble tower, sited close to a busy marketplace at the foot of the hill of the famous Acropolis, rises forty-two feet into the air and measures twenty-six feet across, and it was an astonishing sight for the people of this crowded and vibrant city. The external walls were covered in brightly colored reliefs and moldings representing the eight winds, with each of the eight walls, and a semi-circular annex, carrying a sundial. Inside, the ceiling was painted a stunning blue color covered with golden stars. At the center of the imposing interior was a water clock, which was fed from a sacred source high up on the hill of the Acropolis called the Clepsydra, a name which became synonymous with all water clocks. The clock is believed once to have driven a complex mechanical model of the heavens themselves, like a planetarium, orrery or armillary sphere.

Nobody is quite sure when the Tower of the Winds was built, but it was probably about 140 BCE. As with the sundial at the Roman Forum,

Tower of the Winds, Athens, photographed in the twentieth century

we can think of it as an early public clock tower, giving Athenians the time of day as they went about their daily business at the market and elsewhere, and giving order to their lives. It was also symbolic of a wider order. The gods of the winds, depicted on its decorative panels, were allegories of world order; the stars inside, together with the water clock and its mechanical replica of the heavens, were symbolic of a cosmic order. Certainly, it was an astonishing spectacle.

But, also like the sundial proudly installed by Valerius in Rome, the Tower of the Winds may have carried a further message. If, as some historians believe, the structure was built by Attalos II, king of the Greek city of Pergamon, to commemorate the Athenian defeat of the Persian

Navy in 480 BCE, then it could serve as a vivid peacetime reminder of the military strength of the state—and the discipline needed to maintain it.

Even more historically sketchy is the tantalizing possibility that the city of Verona, once part of the Roman Empire but by the year 507 ruled by the Gothic king, Theodoric, contained a tower that housed a huge water clock, set right by the Sun, which not only showed the time but sounded it in an extravaganza of noise. A scholar working in Theodoric's court explained that "musical instruments sound with strange voices obtained by the violent springing up of waters from beneath," and it is hard to imagine a more potent expression of the power of the new Goth order in the city.[3] Theodoric himself explained the purpose of the clock: to let the people of Verona "distinguish the various hours of the day and thus decide how best to occupy every moment."[4] A sundial on a high tower might be missed or misread. With a monumental acoustic clock tower calling out the hours just outside Verona's city walls, time, and the order it implied, could not be ignored.

IN EMPIRES AROUND the world, the sight and sound of time from high towers had begun to organize the lives of the people, and project a message of power and order. Long before Verona's acoustic clock tower was built, towers carrying drums or bells loomed large over imperial Chinese towns and cities, often centered on marketplaces. The second-century Chinese scholar Cai Yong explained, "When the night clepsydra runs out, the drum is beaten and people get up. When the day clepsydra runs out, the bell is struck and people go to rest."[5] A third-century description of the tower built over the marketplace at the ancient city of Luoyang read, "A drum was hung in the building. When it sounded, the market was closed. There was also a bell. When it was struck, the sound was heard within fifty Chinese miles."[6]

Over 1,000 years later, in the late thirteenth century, when the Venetian merchant Marco Polo visited Kublai Khan's capital city of Dadu (in today's Beijing), he found two towers standing tall over the city center. One tower contained drums, the other a bell, and all were sounded every evening to mark the start of a strict curfew as measured by a water clock. Anyone caught on the streets after the curfew sounded would be arrested and beaten by troops patrolling the city through the night on horseback. Across imperial Japan, too, from at least the eighth century onward, each major city, whether a capital such as Nara or Kyoto, or a more distant outpost, had its clepsydra and a tall tower from which time was sounded to the public, as well as raising the alarm when the population was in danger from fire or attack. Clock towers were part of the ordering infrastructure of cities.

———

IT IS TEMPTING, in the twenty-first century, to feel that we are the first generation to resent being governed by the clock as we go about our daily lives; that we are no longer in control of what we do and when we do it because we must follow the clock's orders. During our long warehouse shifts, sitting at our factory workstations, or enduring seemingly never-ending meetings at the office, we might grumble that the morning is dragging on, but we cannot eat because the clock has not yet got around to lunchtime. But these feelings are nothing new. In fact, while the public sundial was new to Romans in 263 BCE, it had been in widespread use long before that in other cities around the world; the first water clocks date back even further than sundials, more than 3,500 years to ancient Babylon and Egypt.

Public time has been on the march for thousands of years. It is easy to think that public clocks are an inevitable feature of our lives. But by looking more closely at their history, we can understand better what they used to *mean*—and why they were built in the first place. Because wherever we are, as far back as we care to look, we can find that monumental timekeepers mounted high up on towers or public buildings have been put there to keep us in order, in a world of violent disorder.

THE BEST WAY to reach the ancient Italian canal-city of Chioggia is by fast motorboat, which is how I arrived, with a group of clock special- ists, on a wet February morning in 2018. This island enclave, sitting in the Lagoon of Venice fifteen miles south of its better-known sib- ling, thrived for centuries on its salt and fishing industries and as a commercial port, and today tourists join the fishers around the city's docks. Even on a cold and blustery winter's day, Chioggia radiates a picturesque beauty, and it can be hard to imagine the city as a medieval technological crucible.

Our destination that morning was the bell tower of St. Andrew's Church, on the main Corso del Popolo piazza running through the heart of the historical old city, and our guides for the day were Marisa Addo- mine, an engineer and clock historian, and her husband, Daniele Pons. What we had come to see was a mechanical device that had first been set running more than 600 years previously, and our excitement could barely be contained. The object we were about to visit, housed high up in the tower, is the oldest mechanical clock in the world.

The residents and government of Chioggia are rightfully proud of their horological heritage. We were met off our motorboat on the harbor front by the city's mayor and the council's head of culture for an official welcome ceremony, before heading to the local primary school, where the students performed for us a lavish musical pageant based on the his- tory of the clock in the nearby tower. It is a big part of their local identity, and it is easy to see why. When Chioggia's historic clock was set going, in early 1386, mechanical clocks, comprising intermeshing gear wheels driven by falling weights, had been in existence only for about a cen- tury. None older than the clock at Chioggia are known to have survived, though another clock dated to 1386 exists in Salisbury Cathedral in the UK. This is a remarkable enough claim to fame in any case, but there is a sense of local rivalries in this enclave city, too. As Addomine has remarked, "For centuries the Cinderella of the Lagoon, Chioggia now can show something that Venice does not have: a medieval clock."[7]

But Chioggia's claim to technological fame does not rest solely with this clock. The city had also been home to two of the world's most celebrated medieval clockmakers: Jacopo de Dondi, who constructed a remarkable astronomical clock in nearby Padua, installed in 1344, and his son Giovanni de Dondi, born in Chioggia, whose clockwork planetarium, finished in 1364, has thrilled scholars and collectors ever since. Perhaps Giovanni had a hand in the construction of the clock now in the Chioggia bell tower; or, more likely, he and his father had helped build a culture of clockmaking in the city which made it a magnet for mechanical innovators keen to show off the latest technology to the public.

The mechanism of the clock itself, housed in a small room near the top of the bell tower, is imposing but not especially grand. It stands about a yard and a half high, a set of wheels, pinions, barrels and levers all held within an iron frame that is painted red. Rods lead to the hands on the public dial outside, high up on the tower; the clock's bell is struck by hammers pulled by wires. But it was not built as a church clock, calling the faithful to prayer. It arrived at St. Andrew's only in 1822 after it had spent more than four centuries as the public clock mounted on a tower of Chioggia's city hall that used to stand a couple of hundred yards down the Corso del Popolo, before its nineteenth-century demolition.

In the early 1380s, Chioggia, once a great and prosperous city, had been utterly devastated. First, the Black Death pandemic had reached the city in the spring of 1348 and spread quickly, killing half the population and shattering Chioggia's economy and trading networks. Then, in 1379, a grueling, centuries-long conflict for economic supremacy between the rival maritime republics of Genoa and Venice erupted into months of bloody warfare in Chioggia, ending in 1380 only after the death of some 3,600 of Chioggia's population, and ruin for the lagoon city.

An eyewitness described the aftermath of just one battle, a fight that had reached the piazza near the city hall: "there was great destruction . . . the Piazza was stained red with the quantity of blood of so many Christians, killed in a grievous and cruel massacre."[8] Hundreds of bodies lay dead in the street on that one day alone, killed by fire and sword; months of further fighting and siege followed. Supplies were cut off and some residents resorted to eating dogs, cats and even rats to survive. Chioggia's war cast a long shadow over the city and, as it lay in ruins, the task facing its government was the reconstruction of a once-thriving economy and the restoration of Chioggia's status as one of Europe's great cities.

What makes the installation of a public clock on the city hall in 1386 so remarkable is that it took place amid huge cutbacks in public spending in the aftermath of the war. Staffing levels at the city hall, which incorporated the city council offices, court and jail, were slashed almost by half, with the loss of crucial posts, including medics, legal staff and trading-standards inspectors. All the money saved went directly toward the rebuilding of the devastated city center and its infrastructure, from sea defenses, forts and mills to salt beds, housing and the lime kilns needed to support large-scale construction. It might therefore seem odd, given the austerity of spending in the city, that on February 26, 1386, Chioggia's council gathered in the main chamber of the city hall to record the completion of work on the clock, approve the final payment to the clockmaker, Pietro Boça, and agree to pay him a salary of five pounds per month for its ongoing maintenance.

We do not know whether the clock was brand-new or secondhand, or even whether the payment to Boça was for a major refurbishment of a damaged existing clock. What we do know is that getting a working public clock on the tower of Chioggia's city hall, overlooking the main piazza, was considered so important that it was worthy of great expense, just six years after the war had ended. In the midst of so much devastation, with the local economy on its knees, the horrors of a brutal occupation a painful recent memory, and the effects of the Black Death pandemic still being felt in every family across the city, what would a clock bring to Chioggia's city center, as it was rebuilt, phoenix-like, from ruin?

The piazza—today's Corso del Popolo—held a crucial place in Chioggia's heart, as it does to this day. I could *feel* Chioggia as I walked down its wide central street that wet February morning in 2018. With the city hall at its center, the piazza has always been proudly symbolic of the people and spirit of the city. And, in the fourteenth century, its communal bell, mounted up high, *spoke*. It spoke to, and for, the people. What clocks did was mechanize the striking of this bell, and therefore Chioggia, the city of Jacopo and Giovanni de Dondi, two of history's greatest clockmakers, knew better than most the power that clocks could exert. What could Chioggia's new public clock provide, after so much havoc? As a symbol of rebirth, the clock stood for much-needed stability and order. It stood for Chioggia itself.

———

WHETHER THEY BE looted sundials on triumphal columns in ancient Rome, fantastic animated clepsydras in an Athenian marketplace, drum towers sounding the town curfew in imperial Beijing and Kyoto, strange temporal voices screaming across the streets of Gothic Verona, or bell-ringing mechanical clocks high over the bloodstained piazza of medieval Chioggia, public clock towers have always been used to project political power. By looking at who commissioned them and when, we can see how they became symbolic of political concerns: of civic pride, of local identity and, above all, of a sense of order, particularly in the aftermath of disorder—war, occupation, struggle. By forcing a sense of *temporal* order on to the population, these public timekeepers stood for a wider sense of *civic* order. At least, that is what governments hoped. And, long after clocks and watches became commonplace in workplaces and homes, these public clock towers kept their potency. In fact, we might

even conclude that they became ever more useful in projecting power
and keeping control as time went on.

———

FIVE CENTURIES AFTER the government of Chioggia installed a tower
clock on their city hall in 1386, overlooking the scene of brutal massacres
recently carried out by occupying forces, a new series of violent occupations
had reshaped the global map and forcibly changed the lives of hundreds of
millions of people. By the 1880s, the British Empire included India, Australia, New Zealand and Canada, and the so-called scramble for Africa
was seeing Britain take land and people up and down the African continent—"from Cape Town to Cairo," as the British imperialist Cecil Rhodes
described it in 1892.[9] And clock towers accompanied them on their march.

At Africa's Cape of Good Hope, the first British time signal had begun
to operate in 1806, within weeks or even days of the invading force seizing control. Its symbolism could not be clearer, either to rival imperial
powers or to the indigenous African people who were being systematically and brutally displaced from their land. It was a powerful British
cannon, fired each day at noon from a battery high up on the hill overlooking Table Bay. Before long, it was joined by time-signaling towers at
Cape Town itself and, soon enough, right along the coast.

In southeastern Australia, from the 1820s onward, British colonizers
engaged in an energetic program of clock tower building as they imported
Western ideas of discipline and order. The first towers sprang up in towns
like Windsor, Parramatta and Campbelltown. Then Tasmania joined the
march, ordering six clock mechanisms from a leading London maker
for installation in clock towers across the region. Gold was discovered
in western Victoria in 1851, leading to a rush of settler-colonizers to the
area. Melbourne rapidly became a boomtown in the imperial trade network, and with the colonizers came clocks. Three years after the gold
rush began, Melbourne had become home to some thirty-four clockmakers, whereas in the early 1840s there had been just four, and the air of

this colonial trading port was filled with the compulsive sound of European time. By the late 1880s, more than thirty public buildings across New South Wales had been fitted with large tower clocks by the Sydney clockmaker Angelo Tornaghi. For invading settlers and indigenous people alike, the time of the British Empire in Australia was rarely far from view or earshot. As the historian Giordano Nanni has said, "If the clock was an avatar of Western time, the bell was its amplifier."[10]

But it was in India that the British clock tower project reached its greatest and most zealous heights. Britain's grip on India tightened hard in the late 1850s following a bloody uprising in 1857 against its rule. Originally centered on Delhi, the fighting later moved to Lucknow and beyond, becoming known as the First War of Independence. Hundreds of thousands of people were killed in the conflict, with massacres in numerous cities as the British feared losing their vast territory. In 1858, after the rebellion had been suppressed, direct rule by the British Crown replaced a more complex arrangement which had seen the region controlled by the East India Company, a proxy for the British state. The British Raj, as Britain's Crown rule in India was termed, brought sweeping changes across the region as the incoming colonizers sought to stamp their mark on the lands they now ruled. Building construction formed a crucial part of their program, and tall clock towers, with loud bells chiming the quarters and striking the hours across the region, played a vitally symbolic role, being commissioned for every major Indian town occupied by the Raj. Sited at town centers, on major crossroads and on important public buildings, they were impossible to miss.

In Delhi, seat of the 1857 uprising, incoming British officials were quick to mark their arrival and trumpet their achievements back in the home country:

> The latest improvement is the new clock-tower, which stands in the centre of the Chandnee Chowk, opposite the town-hall . . . This building is erected on an appropriate site at the crossing of four streets, and stands 110 ft. high, exclusive of the gilt vane and

finial . . . The dials of the clock are sufficiently elevated to be visible from the East Indian Railway Station, and from other prominent points in the city.[11]

Lucknow, the second center of the mutiny, also received a huge clock tower, this one rising to nearly 220 feet, more than two-thirds of the height of Big Ben in London, which had itself been completed two years after the 1857 uprising. Built in the mid-1880s, Lucknow's clock tower was described as "the largest in India," with brightly illuminated dials driven by a clock mechanism "of great size and power."[12] Each quarter hour, it sounded the Cambridge chimes across the Indian city. Here, as in Delhi, the newly built European clock towers carried a clear and powerful message: the British are here to stay and will crush anyone who steps out of line.

Many elsewhere were put up to accompany new schools and colleges built by the Raj to teach the children of local leaders and officials. One example was the clock tower built in the 1870s as part of Mayo College, a new boarding school campus at Ajmer, in today's Rajasthan, intended to be the Eton of India. The tower itself was designed in a Western architectural style, in contrast to the mixture of traditional Indian styles comprising the rest of the school. At nearly a hundred feet in height, the clock loomed over the area—visible to anybody approaching Ajmer, long before the city itself came into view.

Ajmer was home to a garrison that provided the British with military control over the region; the new school clock tower represented a further form of order. The historian Sanjay Srivastava has explained:

> If the military garrison stationed in Ajmer represented British political control of the physical territory surrounding the College, then the insertion of the clock-tower into a jumble of diverse architectural styles amounted to a proclamation of the capture of an alien culture; the tower "controlled" the "Oriental architecture" surrounding it.[13]

Approach to Ajmer showing Mayo College with clock tower, photographed c. 1900s

The top of the clock tower even took the form of a huge iron crown, like the one worn by Queen Victoria, who was proclaimed Empress of India in 1877, the year that construction of Mayo College began.

In total, over 100 clock towers were built in India during the colonial period; not all by British colonialists, but nevertheless an astonishing construction program that means we must see these clocks as powerful tools in the armory of a colonizing force intent on suppressing disorder. In Ajmer, and across India, clock towers *were* the Empress of India, standing tall and commanding over her domain.

AS THE NINETEENTH century moved into the twentieth, and global empires continued to shift, clock towers had matured into structures that helped express the power of the government at a distance; they became architectural proxies for the state itself, particularly when large territories were involved. Or, at least, that was what some people thought. From the 1880s to the 1900s, the Ottoman ruler Abdülhamid II, his empire breaking up, encouraged a major program of clock tower construction across the Ottoman Empire. During his reign, dozens of monumental clock towers were built by the central government and local officials across a vast region, including land in today's Turkey, Syria, Lebanon, Israel and Libya. The new structures occupied central city locations or prominent hilltops visible for miles around. Describing one such clock, built in 1882 in Adana (now in southern Turkey), were the words:

> Such a huge masterpiece that none can compare,
> Outwardly, a clock chimes, but in essence the government is calling.[14]

With clock towers, central governments believed they could project their power far across the land, though they could also be ways for provincial leaders and officials to express a connection to (or rivalry with) other parts of the empire. Whatever the circumstances of any individual clock, power politics was never far away.

We cannot ignore the violence that has always been embodied in these clocks, and this is a theme to which we will return throughout this book. Earlier, we learned about the drum and bell towers seen by Marco Polo in what is today's Beijing, built by the Mongol conqueror Kublai Khan in the 1270s to help cement his control over the land he now ruled. The time signals from these towers, and their predecessors, ordered the daily lives of Beijing's residents for centuries. But it was another set of imperial invaders that brought about the destruction of Khan's drum tower and, in a wider sense, ended an ancient tradition of public time-telling in China, as a new set of forces jostled to stamp their authority on the great empire. In 1900, after an alliance of eight Western nations— Austria-Hungary, Britain, France, Germany, Italy, Japan, Russia and the USA—invaded China, the old drum tower in Beijing was seized and its leather drumheads destroyed. Then the Western invaders began building their own towers housing modern mechanical clocks across the city.

Take the clock towers built on Beijing's first railway station and on two of its biggest banks as an example. The sites chosen for these Western structures might have seemed like an exercise in practical efficiency and would not raise an eyebrow in any city today. We are used to seeing clock towers on stations, banking headquarters and civic buildings. But in Beijing these choices carried powerful messages. With three clocks, three sides of the city's symbolic Tiananmen Square were closed in. In the words of the historian Wu Hung, "these towers not only surrounded the most prominent political space in traditional China but also overpowered it."[15] With their height, their complexity, their modernity and their political geography they represented no less than the attempted overthrow of an old order by a new one.

It has ever been thus. The connection between the sundial mounted on a column in the Roman Forum in 263 BCE and the clock towers built by Britain across its imperial possessions in the nineteenth century is not an entirely tenuous one. Ancient Rome was an inspiration for the entire British imperial project. Rome's legal system, its architecture, its colonizing ambition, even its complex urban infrastructures offered thrilling visions to Western colonialists, schooled in the history of the classical world. These were visions of *order*, and from Valerius' rostrum in Rome to the buildings around Tiananmen Square, clocks had always been raised up high in the aftermath of disorder, to deliver a common message: know your place; stay in line; obey your rulers.

2

Faith

Castle Clock, Diyār Bakr, 1206

As book launches go, al-Jazarī's was certainly one of the more ostentatious. In front of the party gathered at the palace that day was the glittering star of the show, the machine that took pride of place at the start of al-Jazarī's great treatise. It was a mechanical clock, termed the "castle clock," driven by water, and offering its viewers the most astonishing spectacle of faith.

The castle clock took the form of a house, "about twice the height of a man" in al-Jazarī's words, with a large central doorway.[1] Above this doorway were two rows of twelve smaller doors, with a single crescent Moon, made of gold, able to move along in front of them. Above that was a brightly colored rotating disc showing the twelve signs of the zodiac, each picked out in "gold and other beautifying colours" and pierced with the pattern of its constituent stars backlit by oil lamps, together with a moving spherical Sun and Moon; to the sides of the doorway were two large brass falcons with their wings outstretched over vases carrying suspended cymbals made of white bronze. Over the doorway's arch were twelve glass roundels, and at ground level, almost life-sized,

Fourteenth-century illustration of al-Jazarī's
castle clock

stood five moving human figures: two drummers, two trumpeters and a
cymbal player.

Al-Jazarī explained what guests would see as the day and night wore
on. At daybreak, the crescent Moon would start to move imperceptibly
in front of the lines of twelve doors, at a speed of one pair of doors per
solar hour. As each pair was passed, the upper door would open to reveal
a human figure (it could be anybody that al-Jazarī chose to represent)
and the lower door would flip round, showing a different color to signify
that the hour had been reached. The two falcons would lean forward,
raising their wings and lowering their tails, each dropping a ball from its

beak into the vase below to strike the cymbal loudly enough to be "heard from afar" before retreating. "This happens at the end of every hour until the sixth," al-Jazarī remarked, "at which time the drummers drum, the trumpeters blow and the cymbalist plays his cymbals for a while." This occurred at the ninth and twelfth hours, too. At the top of the clock, the zodiac disc also rotated throughout the day, showing the degree of the zodiac and the signs that were visible. The Sun rose in the sky and then fell until nightfall; the Moon also moved through the scene, its illuminated sphere rotating as the month went on to show the phase, from new to full and back again.

At night, the clock would be lit with the most dazzling illumination. Instead of doors flipping and opening, each of the glass roundels over the clock's central doorway was slowly filled with light from oil lamps as its hour progressed. Once the first six had been fully illuminated, al-Jazarī explained, "the musicians do their duty as they do during the day, and similarly at the ninth and twelfth hours, this being the last hour of the night, by which time all the roundels are filled with light." The mechanism that made a trumpet sound was particularly ingenious. Then it would be time for the palace's servants to refill the water clock's reservoir and reset all the moving parts ready for the new day.

The year was 1206, and al-Jazarī was an engineer and scholar working for the king, Nāsir al-Dīn, at his palace in Diyār Bakr, in today's Turkey near its borders with Syria and Iraq. Al-Jazarī was well known in al-Dīn's court, having spent twenty-five years creating the most fabulous mechanical marvels for the king's family, and, after having devoted his life to filling the royal palace with ingenious treasures, al-Jazarī had been instructed by his master to set out his experience in a huge, lavishly illustrated publication entitled *The Book of Knowledge of Ingenious Mechanical Devices*. It showed the state of the art in Islamic technology and engineering in the most vivid detail.

The extravaganza of sound, movement, music, color and light in al-Jazarī's castle clock allowed al-Dīn to demonstrate his devotion to the worship of Allah in the most spectacular fashion. Of course, there was an element of competition with this elaborate and expensive clock, too. By commissioning it, the king could show off the extent of his royal patronage of the mechanical arts, and he was not the only ruler to do so. In 1368, a similarly magnificent automaton water clock was shown off in the royal palace at Tlemcen, Algeria, at a public feast on the prophet Muhammad's birthday. Tlemcen's clock put on an exuberant exhibition and, like the Diyār Bakr clock, was designed to impress with the most astonishing displays. It was the talk of its town for forty years.

If it had not been constructed for a great Muslim ruler, al-Jazarī's monumental clock might equally have been built for display at a mosque. A spectacular automaton clock, like the castle clock, had been installed a few years earlier at the Great Mosque of Damascus, Syria, one of the oldest in the world, and stayed in operation there for at least a century and a half. Fragments of similar clocks still survive in the Qarawiyyin and Bū'anāniyya mosques in Fez, Morocco, originally installed in the fourteenth century.

Or perhaps the clock was used for religious instruction. Thirty years after al-Jazarī published his treatise, a device like his castle clock was set up in the entrance hall of the Mustansirīya college in Baghdad, Iraq, a huge madrasa for Islamic studies that was intended to be the most magnificent in the world.

These automaton water clocks, while never commonplace, were nevertheless a fixture in medieval Islamic society, and it is easy to imagine the power they wielded. Day-to-day timekeeping for religious worship was usually done using sundials, astrolabes (flattened representations of the night sky) or the simple bucket-shaped water clocks that had by then been around for at least 2,500 years. But by making monumental astronomical and automaton clocks like al-Jazarī's castle clock, the people who occupied positions of power could offer spectacular visions of faith to the worshippers in their communities, whether in schools, in mosques or for visitors to their palaces. These clocks were meant to represent God's wisdom, dominion and omnipotence on Earth. This divine power was mediated by the clerics and rulers who governed the Islamic world. For their subjects, clocks were a constant reminder of their place in the universe.

———

FOR ALL THE major world religions, clocks and faith go hand in hand. In the Islamic faith, there are two temporal rhythms that dictate everyday life. The first is the lunar calendar, which is used to define the start and end of Islamic months and the time to make a pilgrimage to Mecca. The Qur'ān explains that the new moons are "but signs to mark fixed periods of time in the affairs of men, and for the pilgrimage." The first sighting of the crescent moon for the fasting month of Ramadan remains one of the most significant temporal observations for Muslims around the world today. The second pattern of times governing the Islamic faith is the regulation of the five daily prayers, which follow a solar pattern, not the lunar observations of the Islamic months. The prayers of *Salat al-maghrib* (sunset), *Salat al-isha* (evening), *Salat al-fadjr* (daybreak), *Salat al-zuhr* (noon) and *Salat al-asr* (afternoon) are all timed by the Sun.

In the Christian calendar, the date of Easter and other moveable feasts requires complex calendrical and astronomical systems that bring lunar and solar cycles together. The daily routine of prayer in Christian monasteries followed a fixed pattern—something like *Lauds, Prime, Terce, Sext, None, Vespers* and *Compline*, although the details shifted over time. Bell-sounding water clocks, or clepsydras, were being used in monasteries at least as far back as the fifth century, and by the ninth century they had become commonplace. The monk known as Hildemar of Corbie, writing near Milan in about 845 CE, offered stern advice to colleagues who were obliged to observe prayer times: "He who wishes to do this properly must have *horologium aquæ*"—a water clock.[2]

Christian practice had drawn in turn from the Jewish system, which divided daylight into twelve hours marked by religious observances, from *Alot Hashachar* at daybreak to *Tzeit Hakochavim* at nightfall, with *Chatzot Halaylah* occurring at midnight.

The Jewish mystical text the *Zohar*, purporting to be ancient but probably written in the thirteenth century, includes a revealing passage on the role of clocks in Jewish religious practice as it describes a journey from what is now the Israeli city of Tiberias:

> Rabbi Abba set out from Tiberias to go to the house of his father-in-law. With him was his son Rabbi Jacob. When they arrived at Kfar Tarsha, they stopped to spend the night. Rabbi Abba inquired of his host: Have you a cock there? The host said: Why? Said Rabbi Abba: I wish to rise exactly at midnight. The host replied: A cock is not needed. I have in my house a device [to announce midnight]. By my bed there is a scale [with a clepsydra hanging from one side which] I fill with water. It drips out drop by drop, until just at midnight it is all out, and then this weight moves, and produces a sound which is audible in the whole house, and then it is exactly midnight. This clock I made for a certain old man who was in the habit of getting up each night at midnight to study Torah. To this Rabbi Abba said: Blessed be God for guiding me here.[3]

In the Sikh religion, prayers are offered during an early-morning period called *Amrit Velā*, at dusk and at nighttime. In Hinduism, time *is* God. In Buddhism, time only exists in our minds—and perhaps our nostrils, as incense-burning clocks were used to mark the passage of time in Buddhist ceremonies in Korea, Japan and China for hundreds of years.

———

WHILE SUNDIALS, WATER clocks, lunar sightings and the shrill morning cries of cockerels had served the timekeeping needs of religious life for centuries, they each had their limitations. Most obviously, these were practical: the Sun did not always shine, and the Moon often hid its face; water clocks clogged up and needed constant attention and maintenance. Incense clocks relied on a costly consumable product. But there was an intellectual limitation inherent in these clocks, too. Astrolabes, which were hand-turned facsimiles of the night sky, demonstrated the complexity of the universe but not the hand of its maker. There was no divine motive force to keep the astrolabe turning in time with the rotations of the heavenly bodies themselves. They *looked* like the universe but did not *act* like it. But in the late thirteenth century, a new type of clock was developed, and this one offered something new.

The mechanical clock, developed in Europe, took the form of an intermeshing set of geared wheels driven by weights falling under gravity, which was regulated so as to rotate at a constant speed (or near enough constant), with some kind of interface that allowed its owners to read the time, either by sight or sound. These new mechanical clocks would run night and day for as long as they were kept wound up and corrected, and what this did in practice was to mimic the rotation of the Earth itself—to keep pace with the Sun's apparent movement across the sky each day.

There have been arguments for centuries about why the mechanical clock was developed in the late thirteenth century, but we might find answers if we stop thinking about the invention of "the" mechanical clock and, instead, imagine two different inventions. One type of clock was relatively simple and answered a practical problem. These clocks mechanized the sounding of alarms for prayer times in Christian monasteries, or communal bells in towns and cities as in the Chioggia clock discussed in the previous chapter. Water clocks had been doing this job reasonably well for centuries but, as we saw, they had practical limitations. The geared mechanical clock was an upgrade.

But the other type of clock, which has become known as the astronomical clock, was something altogether more complex. One, the planetarium clock completed by the Chioggia-born Giovanni de Dondi in 1364, included a perpetual calendar showing not only the fixed religious feasts of Christianity but the moveable ones, too. It also showed the motions of the Sun, Moon and planets, and even displayed movements of the

Moon's orbit that take over eighteen years to complete. By considering these rarer, more complex clocks, we come to the other theory as to why mechanical clocks came about. People were trying to make mechanical versions of the universe itself, with the rotation of the gears representing the rotation not just of the Earth but of all the celestial bodies as seen from it. They were attempting to re-create the force that moved the heavens.

Writing in 1271 for his students at the University of Montpellier, France, the astronomer Robertus Anglicus was considering the rotation of the celestial equator (really the rotation of the Earth itself, though that was not how people saw things then), and was excited by new developments taking place across Europe:

> clockmakers are trying to make a wheel which will make one complete revolution for every one of the equinoctial circle, but they cannot quite perfect their work. But if they could, it would be a really accurate clock and worth more than [the] astrolabe or other astronomical instrument for reckoning the hours.[4]

The idea behind this thinking was that the universe was God's prototype. And the question being asked was this: could humans make production versions of that prototype? Such mechanisms would need not only to represent the motions of the heavenly bodies—accurately and precisely—but also to run continuously, at the same speed.

This brings us to what geared mechanical clocks might do that water clocks could not. They held out the promise of precision motion, at the constant speed of the universe, that would run forever. They were production versions of God's perfect prototype, and what a prize it would be for the clockmakers who succeeded in building them. They could be the creators of their own universes. And as far as we know, it was just four or five years after Robertus Anglicus wrote his hopeful remarks that the first successful mechanical clocks had, indeed, been made.

———·———

AT THE HEART of the desire for clocks to mimic the motions of astronomy was astrology, the belief that the movement of the heavens affects
the lives of people on Earth, and this was of the most profound importance to medieval European society. *Everybody* believed that the stars
influenced affairs on Earth, just as everybody believed in a god. Clocks
showing the movements of the heavenly bodies were there to be *obeyed*.
And it made a lot of sense. "If the Moon can affect the tides, why should
the stars and planets not affect the natural world in their own ways,"
asks the historian John North.[5] When de Dondi's clockwork planetarium
ended up at the court of the Duke of Milan in the mid-fifteenth century,
its astrological indications became the focus of political obsession. For
medieval Europeans, astrology was a matter of sincerely held faith.

It was the scientific revolution of the seventeenth century that turned
people's minds firmly toward the idea that the universe itself was clockwork, built by a clockmaker God. Robert Boyle was the chemist and natural philosopher who helped usher in the modern scientific method, and
it was Boyle who, in 1686, wrote that the universe was:

> like a rare clock, such as may be that at Strasbourg, where all things
> are so skilfully contrived that the engine being once set a-moving,
> all things proceed according to the artificer's first design, and the
> motions of the little statues that at such hours perform these or
> those things do not require (like those of puppets) the peculiar
> interposing of the artificer or any intelligent agent employed by
> him, but perform their functions upon particular occasions by vir-
> tue of the general and primitive contrivance of the whole engine.[6]

The huge astronomical and automaton clock at Strasbourg Cathedral
made in 1574 offered an astonishing vision of religious teaching and astrological prediction to the masses who crowded in to worship: a celestial

globe, calendars, a device to calculate the date of Easter, automaton figures for each day of the week, and paintings depicting Christian religious
scenes, including the creation of the Universe, original sin, redemption,
resurrection and the Last Judgment. An angel turned an hourglass over.
The positions of the signs of the zodiac were indicated and, once each
day, the four ages of life passed before death. Death itself was a grotesque
skeletal statue which signaled the strokes of the hour using a human leg
bone as a bell hammer. A realistic and finely detailed automaton cockerel
flapped its wings and crowed noisily to warn of the denial of Peter.

It must have been difficult, for devout Christians, to watch this awe-
some daily spectacle and not feel an overwhelming sense of their god's
power and presence on Earth, and the influence of the heavens on their
life. And that is the point. Worshippers could watch the movement of
the heavens and the performance of stories from their religious faith,
and they could *believe*. So, which European city would not want the very
latest religious technology? Which authorities would not want to present
their congregations with such a vivid display of God's perfect universe?
Their power rested on the obedience of their flock: they were all looking
for new ways to get believers through the doors of their churches to sit
in astonishment and wonder. The commandments of the heavens were
to be obeyed, so spectacular clocks sprang up around the world.

———————

BETWEEN THE INTRODUCTION of the first mechanical clocks in about
1275 and the year 1600, nearly 200 towns and cities across Europe, from
Aberdeen to Zurich and everywhere in between, had been fitted with
a public clock. Many were simple devices showing the time and striking the hours on bells, but some were the more complex astronomical
clocks, often with the sort of automaton figures found on the Strasbourg clock. Let us visit just two of these mechanical marvels.

Lübeck, on Germany's Baltic Sea coast to the northeast of Hamburg,
had grown rich on the seaborne trade of northern Europe since it had been

established in the year 1159. Its enormous St. Mary's Church sits in a commanding position over the city. Constructed between 1250 and 1350, it was a towering expression of the power, status and prosperity of the Hanseatic League of cities, to which Lübeck was central. It projected the religious faith of the rising merchant class on an epic Gothic scale.

The church authorities installed a clock behind the altar possibly as early as 1384, although reliable records give us a date of 1405, when the clock was rebuilt. This first clock included a complex calendar dial but was otherwise quite simple; it was badly damaged by two fires over the following century. In 1561, a new clock was commissioned that was an altogether more elaborate matter.

At its base, niches held carved figures representing the deadly sins of anger, lust, greed, unchastity, pride and indolence. Nearby, an inscription exclaimed:

> When Thou regardest in the Heavens the Shining Sun and then the light of the Moon, shining in their appointed courses; so then Thou canst perceive with the eye how the hours fly and hasten away, permitting themselves no delay. So often as Thou hearest the melody of the sonorous bell, think then of God who governs the Stars; and at the same time praise Him.[7]

Above the figures of sin was the huge calendar dial from the original clock, updated for the sixteenth century. It gave the day of the week, the Moon's age, festival days, the time of sunrise, the date, the dominical letter (helping work out the dates of future Sundays), the golden number (helping establish the year's position in the lunar cycle), the year of the solar cycle, the date of Easter Sunday, and the number of days between Christmas and Shrove Tuesday for that year. The center of the dial held tables of lunar and solar eclipses visible from Lübeck, and at the corners of its frame were images of the Evangelists: Matthew, Mark, Luke and John. Signs of the zodiac surrounded the calendar in gold.

Then came the astrolabe dial, or planetarium, which was a dazzling

Lübeck astronomical clock, photographed c. 1870

and complex array of astronomical indications—the Sun, Moon and planets, the signs of the zodiac, all rotating slowly against their pointers. At the center, representing the earth, was the figure of Christ, surrounded by clouds and rays.

Above the planetarium was the *pièce de résistance* of the Lübeck clock, which towered over the heads of parishioners. Beneath a canopy stood Christ holding the Earth in his hand. Over that was a bell tower which held the clock's hour bell. To its right, an automaton figure of Father Time held a bell hammer and an hourglass; to its left was a moving figure holding a torch and skull to show the fleeting nature of time. Above the tower was a statue of the two-faced god Janus, behind which stood a fourteen-bell carillon. Nearby were figures representing the Sun, Moon,

Apollo, Diana, Mercury, Mars, Jupiter and Saturn. As if this was not impressive enough, after the noon bells had stopped striking, a parade of automaton figures emerged, one after the other, each being blessed as they passed in front of Christ. An usher bowed as each figure returned to the innards of the clock. Two angels then lifted trumpets to their lips and a fanfare sounded while music played throughout.

In the early hours of March 29, 1942—Palm Sunday—the church and its contents were destroyed by British bomber aircraft. John Castle, the horological historian who recorded the history of the clock in 1951, concluded, "There are not suitable words available to express the regret at its loss."[8]

Conflict in an earlier period of history surrounded the second clock on our itinerary. Tourists visiting Prague, capital of today's Czech Republic, make a beeline for the astronomical clock fitted to the south wall of the Old Town Hall. This astonishing mechanical creation, with details picked out in shimmering gold and vivid blue, towers three stories high over the streets. It was first built in about 1402 by the clockmaker Mikuláš of Kadaň and the astronomer Jan Šindel, at a time of violent religious upheaval in the prosperous Bohemian city, one of the largest in Europe and seat of the Holy Roman Emperor.

What became known as the Bohemian Reformation had begun in the mid-fourteenth century as a backlash against the authority of the Pope and the Roman Catholic Church. In Easter 1389, the Prague clergy incited a pogrom against the Jewish community living in the city; 3,000 Jews were killed, and their homes were ransacked and burned to the ground. Religious unrest continued as, in 1398, the reforming theologian Jan Hus took up a teaching role at Prague's university, founded fifty years previously and located a few moments' walk from the Old Town Hall. Hus spent the next seventeen years preaching against the corruption and excess of the Catholic clergy until he was executed by burning at the stake for heresy in 1415, leading to twenty-five years of armed conflict between "Hussites" and forces loyal to the Catholic Church.

In reality, church and state power were as one, so the role of the

Prague city council—sitting at the town hall—in this period of intense religious strife must not be underestimated. Prague was at the heart of Catholicism and its mayor and councillors knew they were at grave risk if the revolution fomented by Hus was to take hold. The installation of an elaborate astronomical clock in the most visible public place—the external wall of the town hall, not the interior of a church or cathedral—was a forceful statement of establishment power over Prague's residents, who were unsure of the direction that their city and religion were taking.

In 1410, as Hus's teachings were gathering momentum, the Prague councillors commissioned a refurbishment of the original clock from Mikuláš, who, as they described in a letter that year:

> made the astrolabe in which the sun follows its true course in the zodiac and twelve signs, as in the sky, show there how the sun rises and sets every day, that there is also a sphere which indicates a sign for every day as well as showing when the moon waxes or wanes every four weeks, as in the sky, and written there on the front wheel are all the saints' days for the whole year.[9]

It may not be fanciful to suppose that the installation of this huge clockwork astrolabe, turning in time with the heavens in order to show the "true course," was a clear, symbolic statement to the people of Prague from its establishment leaders: let the universe run in the orderly way it always has, and do not question the authority of those higher than you.

———

SPECTACULAR ASTRONOMICAL CLOCKS that acted like mechanical versions of the heavens, asserting God's power and the complexity of the universe, were, of course, unusual. So, too, were the large bell-ringing clocks set up at the top of towers in our towns and cities. These public clocks formed a backdrop to our increasingly clock-oriented lives, but our most intimate relationship with clock-time has come to be formed with the clocks that we have encountered in our workplaces and homes, around our necks, in pockets and pinned to our clothing.

Small clocks, driven by falling weights or by tightly coiled springs, began entering our everyday lives from the fifteenth century, and watches, portable timekeepers we can carry about our person, started to appear from the early sixteenth century. Of course, it took a long time for these high-tech instruments to become commonplace, or, rather, evenly distributed. Class, gender, geography and a host of other factors affected the likelihood any individual might come to live by the clock. But as time passed on, and these devices wormed their way through societies, our relationship with the time they told us changed. Awareness of clock time was ever present, and personal. And a new idea grew up alongside them. With clocks always in view, we started buying into the idea that time could be wasted.

This concept, growing in significance for some time, really took off when the English Puritans of the sixteenth and seventeenth centuries started pushing the idea of a work ethic as the foundation of pious devotion. The seventeenth-century Puritan pastor and theologian Richard Baxter was uncompromising in his teachings on the subject. In an essay entitled "The Redemption of Time," he asked, "If you idle away this life, will God ever give you another here? If you do not work well, shall you ever come again to mend it?" In *A Treatise of Self-Denial*, he claimed that idleness was a most heinous sin, for by wasting time, "you are guilty of robbing God himself. It is him that you owe your labours to; and idleness is unfaithfulness to the God of heaven that setteth you on work: ever in working for men, you must do it ultimately for God."[10]

As it took hold in the minds of the faithful, the idea started to become embedded in the hardware of timekeeping. A new style of pocket watch was developed, known as the "Puritan" watch, which was undecorated and austere—a deliberate contrast to the exuberant and highly decorated watches that were then flooding the aristocratic marketplace. What better than one of these plain and humble devices to remind you of the sin of sloth? The first watch ever to be acquired by the British Museum, in 1786, was a Puritan watch, made in about 1635 and alleged to have belonged to Oliver Cromwell, who considered himself a Puritan Moses.

These everyday items provided a form of disembodied discipline for the Protestant work ethic, as Lewis Mumford, the twentieth-century chronicler of modernity, pointed out: "Time-keeping passed into time-serving and time-accounting and time-rationing."[11] It was clocks and watches, as much as sermonizing pastors, that disciplined the masses with a message of pious servitude.

Personal, domestic and workplace clocks and watches, which we invited into our lives in ever-greater numbers, became our overseers and landlords, our managers and supervisors. And this was an expression of faith as much as capitalism. We still live under the spell. Today, we live in a world that equates personal timeliness with moral righteousness more than ever before.

The phrase "time is money" became famous after it was used by the American politician and scientist Benjamin Franklin when, in 1748, he wrote a guide offering "Advice to a Young Tradesman":

> Remember that Time is Money. He that can earn Ten Shillings a Day by his Labour, and goes abroad, or sits idle one half of that Day, tho' he spends but Sixpence during his Diversion or Idleness, ought not to reckon That the only Expence; he has really spent or rather thrown away Five Shillings besides.[12]

It had been a long time in the development, but the secular idea set out by Franklin drew deeply from the sacred notion that time was God's time, an idea preached by early Protestants just when personal clocks and watches were becoming part of everyday life for the middle and upper classes. It soon filtered down to the rest of us, and today there are few in the West who do not subscribe to the time-is-money doctrine.

THERE WE ARE. Two extremes. On the one hand, we have the myriad mundane clocks and watches that surround us today, exhorting us to work for God (or Mammon) and not to waste the time they tell us. On the other hand, for centuries we have gasped in awe and wonder at monumental astronomical and automaton clocks in mosques and churches and on public buildings: mechanical simulacra of God's perfect universe. This, too, is a story that comes right up to the present day.

King Abdullah University of Science and Technology sits on Saudi Arabia's Red Sea coast near Mecca, looking over the vast expanse of blue-green water toward Sudan and Egypt. I reached the university campus by a one-hour drive in a gleaming Chevrolet Suburban SUV in the dead of night, passing through the desert-like plains surrounding Jeddah, some of the hottest lands on Earth. As we drove along the almost deserted highway, I was struck by how brightly the Moon, which was almost full that night, was shining; living in London I had forgotten how clear and crisp the heavenly bodies could appear in a dark sky. Eventually we turned off the main road from Jeddah at the village of Thuwal and the driver started negotiating the security checkpoints surrounding the university. As we drew closer to the campus's heart, the scale of the university became clear and impressive: at fourteen square miles it occupies an area twelve times the size of the City of London.

The university was founded in 2009 as part of Saudi Arabia's twenty-first-century focus on scientific and technological innovation, and I was there to speak at an international conference on the science and technology of time. I found myself surrounded by experts from around the globe offering rich insights into aspects of the subject I had never encountered. Time is at the heart of so many disciplines that it made the perfect subject for an interdisciplinary conference.

The focus of the meeting was on cutting-edge research, but the Museum of Science and Technology in Islam, at the heart of the campus, shows that innovation in the region is nothing new. From the seventh century onward, the Islamic civilization was a crucible of new inventions, ideas, theories and manufacturing methods, and the museum describes vividly the diversity of Islamic contributions to today's science and technology, from mathematics, chemistry and biology to architecture, astronomy, navigation and mechanical engineering.

Once I had delivered my presentations, the museum's leader, Ana Catarina Matias, knowing I was a curator as well as a time historian, arranged to have me personally escorted round the museum's exhibits. There was one display I particularly wanted to see. I had recently read the English translation of al-Jazarī's *Book of Knowledge of Ingenious Mechanical Devices* from 1206 and had been captivated by his description of the castle clock, lost long ago. And here, standing in a museum near Mecca, the most holy site in Islam, was a faithful working replica of al-Jazarī's clock, with all its bright color, moving figures and brilliant illumination.

It was a treat for the senses, and certainly brought me closer to the work of the medieval Islamic clockmakers whose mechanical marvels had so thrilled and awed the people who saw them in schools, mosques and palaces. But there is nothing quite like seeing original artifacts, not reconstructions. The university campus is about seventy-five miles from the Great Mosque of Mecca, with its central cube-shaped Kaaba, the most holy site in the Islamic faith. The clock I *really* wanted to see during my visit to the country sits at the top of the Makkah Royal Clock Tower, part of the government-owned Abraj Al-Bait skyscraper hotel complex overlooking the mosque site, and is the largest clock in the world. It is also, at more than a third of a mile above street level, the highest. But access to the holy city is prohibited to non-Muslims, so I was forbidden from visiting.

What I do know is that the Makkah clock, fully completed in 2012, looks remarkably like London's Big Ben, albeit much bigger—each dial on the Makkah clock is 141 feet in diameter, compared with Big Ben's 23 feet—and designed in an Islamic architectural style. I also know that, when the clock was completed, Makkah officials lobbied to get the

Makkah Royal Clock Tower overlooking the Great Mosque of Mecca,
photographed in 2012

world's prime meridian, currently running through Greenwich, changed to the meridian of Mecca. "Putting Mecca time in the face of Greenwich Mean Time, this is the goal," said Mohammed Al-Arkubi, the hotel's general manager, in the Saudi press.[13]

It is not an astronomical clock, but the tower houses an astronomy exhibition behind the clock dials, and there is a lunar observatory above the clock room, used to sight the Moon to mark the start of holy months. The call to prayer, or *azan*, flashes five times daily from the clock tower's 21,000 white and green lamps, visible for nineteen miles.

The whole complex was built on the flattened remains of an eighteenth-century fortress, originally constructed to protect Mecca from invasion, but bulldozed together with the very hill it sat on in the early 2000s to make way for the hotel complex.

Mecca is the religious heart of Islam and, as the scholar and critic Ziauddin Sardar has written, "What happens to and in the city has a profound effect on Muslims elsewhere."[14] For centuries, we have gazed up at monumental clocks in mosques and churches with a sense of awe. As we have done so, these clocks have strengthened the authority of our religious leaders. In that sense, the modern Makkah clock is simply the latest in a line of devices performing this role stretching back a thousand years. It is a message to the Islamic world of the power and status of the clerics in Saudi Arabia. But it is also, in a post-colonial world, a message to the West. Clocks like Big Ben embody national identities. The construction of a clock that looks like an Islamic Big Ben, but six times the size and a third of a mile tall, reminds us that fortunes wax and wane; the wealth and power of nations rise and fall. Big Ben is from a past age. Who rules the world now?

3

Virtue

The Hourglass of Temperance, Siena, 1338

Ambrogio Lorenzetti gazed out of the window of Siena's Palazzo Pubblico, the city hall, for artistic inspiration. The year was 1338, and the view that faced him was of a divided and troubled city. On one hand, the evidence of prosperity and innovation was everywhere to be seen. There were institutions of banking, financial services and commerce. Venture capitalists were investing heavily in Siena's businesses and residents, and profits from trade connections as far afield as China were pouring into the city. Masons were busily constructing buildings to house Siena's growing population. Outside the city walls, carefully tended fields and gardens provided food and materials.

But Lorenzetti could see danger, too. Siena was a self-governing republic run by the *Signori Nove*, or Nine, a council of banking and merchant oligarchs. The city was embroiled in a bitter struggle with the neighboring republics of Florence and Pisa, with war an ever-present threat. This was part of a wider battle between the Roman Catholic Church, represented by the Pope, and the emperors of the Holy Roman Empire, based in northern Europe. Sometimes, this political battle spilled over into violent conflict, with cities fighting cities, and with deep bloody rifts opening within cities and even among families. But at greater risk was the idea of republican government itself in the face of those who wanted to sweep it aside and vest power in a single, autocratic ruler, whether a monarch or a dictator. At stake was the right to control land and people across Europe, and war was never far away. Alliances and factions constantly shifted, and for politicians in the wrong place at the wrong time it could mean life or death. In fourteenth-century Siena, some of the greatest political ideas in history were playing out in the meeting rooms of the city hall.

With this intense political background in mind, Lorenzetti turned his attention to the Palazzo's Council Chamber in which he was standing. He had been commissioned by the Nine to cover the walls of their meeting hall with a series of wall paintings, or frescoes, that told a story of political virtue. What they wanted laid out for all to see was a depiction of Siena's prosperity under their own government, and a contrasting vision of what would happen if the people decided instead to invest their support in a single, autocratic ruler. This was not just an abstract threat: autocratic local rule was sweeping Italy, and Siena had suffered under it before. What the Nine demanded was a detailed and powerful depiction of good and evil; of peace and of war. This was to be a bold work of propaganda: a vivid description of the Nine's political program, and one that was grounded in the daily experiences of Siena's citizens. The Nine were sending a powerful message to the people of Siena, as well as to their own colleagues, and Lorenzetti was the person to carry out their plan.

These brightly colored and exquisitely detailed paintings, completed in 1339, can still be seen at the Palazzo, and a trip there is in every sense a revelation. I visited with a group of friends in 2019. We chose to make our excursion in the middle of winter to avoid the crowds of summer tourists. Even so, the streets were full, and we found the building a welcome respite from the bustle of the city's central Piazza del Campo outside. Inside the Palazzo, the rooms were cool and quiet, and we reached the chamber in which Lorenzetti's paintings reside, known as the *Sala dei Nove*, by passing through a series of larger halls, each filled with the decorative treasures of a city that was once one of the richest in the world. Finally, passing through a low doorway, we had revealed to us the sequence of dramatic and thrilling images. They were sumptuous visions stretching right up to the high and ornate ceilings; images that once looked down onto the merchant oligarchs that ruled this commercial metropolis.

I stood with my back to the wall which contains the room's only window and surveyed the images that line the chamber. On the wall to my left, a panoramic landscape laid out Lorenzetti's vision of the effects of war through bad government, showing to the room's visitors the violence, destruction and loss of property they would face under the rule of a tyrant, depicted as a monstrous, horned, cross-eyed devil. Bad government looked out for itself, not the common good. The vices of Avarice, Pride and Vainglory hover menacingly over the tyrant's head. Cruelty, Betrayal and Fraud sit at his right hand; Fury, Division and War at his left. Justice has been bound in chains at their feet. A woman is being assaulted and dragged away by a soldier and a merchant. At their feet lies a man, either dead or dying, his stomach slashed open and his bowels spilling out onto the ground. Buildings are on fire. The city is at war with itself. The countryside, outside the city walls, is barren. Fear pervades the scene.

I turned instead to the wall to my right, directly opposite the images of war and tyranny. Here I saw a longer panorama that set out a vision of peace through good government. In this scene, members of the city community are working in the service of each other for the betterment of all. Buildings are under construction and residents trade freely and cheerfully. There is a wedding procession and people are dancing. Outside the city walls is a rich, lush and productive countryside, laid out carefully and rationally. The two halves of the scene sit in harmony with each other. The figure of Security hovers reassuringly overhead, although by holding in her hand a wooden gibbet containing the dead body of a hanged criminal, she gives viewers a reminder that peace comes at a price. But by paying the price for this good government, Siena prospers in this scene, and everyone enjoys the benefit.

The political vision laid out in these two images was clear. Good government, and the widespread shared prosperity it would bring Siena's citizens, required virtue, not vice. Republicanism, not the tyranny of a local despot from one of the rich families, was the route to a good life. The third fresco showed how it would be done. Occupying the central position between the two landscapes, facing the window of the hall and thus bathed in the brightest light, is the painting that offered a symbolic representation of Siena's ruling Nine themselves.

An imposing bearded figure near the center of the frame stands for the civic government of Siena. Seated nearby is the figure of Peace, holding an olive branch. Above their heads are the three holy virtues of Faith, Charity and Hope. The civic virtue of Justice sits in a commanding position, with a mighty sword in her hand and the decapitated head of an executed criminal in her lap. Seated among them are the remaining four civic virtues of public life: Fortitude, Prudence, Magnanimity and Temperance. Five civic virtues in total, and three holy virtues, and peace itself, all arranged around the personification of Siena's government. Nine virtuous figures celebrating the nine ruling officials who met in that room and who had commissioned Lorenzetti to paint the remarkable series of scenes. As a vivid expression of their manifesto for governing Siena, it was a breathtaking vision of political virtue. It was a blueprint for republican self-government—and the struggle between good and evil. And it was impossible to ignore.

There was one figure in particular that I had come to the Palazzo— indeed, to Siena—to see. She could almost be missed amid the massed figures depicted around the room. Unlike other virtues staring defiantly out toward the viewer, her gaze is averted; she sits toward the right-hand edge of the wall. Yet, despite all this, I spotted her immediately. The virtue of Temperance, with flaming red hair, dressed in startlingly rich pink and cerulean blue robes, holds in her right hand a curious technical instrument, comprising two conical glass vessels joined at their necks, held within a simple wooden frame. In real life it would have measured something like eighteen inches in height and eight inches across, and it is remarkable for the prominence it is given in this crowded painted

Temperance holding an hourglass, painted by
Ambrogio Lorenzetti in 1338

scene. Temperance gestures toward the object with her left hand and
looks down upon it with an expression of concern on her face.

The instrument in Temperance's hand is something that is so familiar
to us today that we only ever think of it as a kitsch plastic novelty that
was once used for timing the cooking of eggs. But in 1338, when Lorenzetti
started painting these scenes, this type of instrument was at the cutting
edge of horological technology, and its inclusion in this Sienese fresco
was about to change the way Western civilization thought of good and
bad, right and wrong, and even the most fundamental human concerns
of all: life and death. Human virtue was at stake, and this timekeeping
device was to become the symbol of a better future. Ambrogio Lorenzetti

had painted what is now the oldest known depiction of an hourglass, and Temperance was warning us that the sands of time were running out.

———•———

NOBODY KNOWS FOR sure when hourglasses were invented. They have a feel of antiquity about them, and some historians argue they were invented in ancient Greece, but the evidence does not stack up. It is more likely that they were introduced about halfway through the Middle Ages. The Islamic scholar al-Bīrūnī was experimenting with using sand instead of water to measure time as early as the ninth or tenth century, though the devices he made might not have been hourglasses. It seems more probable that hourglasses as we know them were developed in the eleventh or twelfth century, by either Islamic or European makers. Maybe hourglasses are not much older than mechanical geared clocks, which were first developed in the thirteenth century.

Why so late in the history of technology, given that simple water clocks had already been around for over 2,500 years by then? On the face of it, sand clocks and water clocks run on the same principle, which involves something in the form of a fluid flowing at a steady rate through a small hole. But there are two important technical differences between the two technologies. Hourglasses need clear blown-glass bulbs, which were harder to make than the carved stone vessels of ancient water clocks, and the techniques to do so were developed later. More crucially, the powder in hourglasses was not straightforward to understand or manufacture and needed to be carefully prepared in order to flow smoothly for the whole duration of the glass, not clog up or stick in the hole between the two vessels. A French recipe written in the 1390s explained the laboriousness of the technique:

> Take the grease which comes from the sawdust of marble when those great tombs of black marble be sawn, then boil it well in

wine like a piece of meat and skim it, and then set it out to dry in the sun; and boil, skim and dry nine times; and thus it will be good.[1]

Once developed, hourglasses found many uses. Ships' navigators used them to time the duration of the watches, and to estimate speed and position; they were sometimes called sea clocks in the early days. Lawyers and politicians used them to time debates and legal arguments. Preachers timed their sermons, and teachers timed their lessons. Monks could time the intervals between offering prayers. In the home they found many domestic uses. In mills and factories, they regulated industrial manufacture. In observatories they counted the time between star transits, and in the hands of doctors they could measure a patient's pulse. They could even be used to time the manual striking of clock bells when the clock itself was too inaccurate or broken. Sometimes, they would run for an hour, which is why they are generally called hourglasses. Most would run for longer or shorter periods of time—from twenty-four hours down to a few seconds—which is why they are sometimes called sandglasses, although this, too, is a misnomer since most sandglasses do not use sand. Egg-timing was the least of their uses, though that is how we most easily remember them today. As timekeepers they have always been cheap and cheerful.

In his 1836 novel *Sartor Resartus*, Thomas Carlyle wrote that "By Symbols, accordingly, is man guided and commanded, made happy, made wretched."[2] To think of hourglasses as simple, mundane machines for tracking the passage of time is to miss their wider importance in history. By studying Ambrogio Lorenzetti's 1338 depiction of Temperance on the wall of Siena's Palazzo Pubblico, we can come to understand the symbolic meaning that would be attached to hourglasses, and it is this symbolism that has given them their greatest impact on civilization. Visual art, in Siena and elsewhere, has always held great power over people. The symbols depicted in art have had real, tangible effects on the lives and minds of the people who have viewed them; they have changed human civilizations. To understand their meaning, we must understand who made these depictions, and why. And, in the case of the hourglass, the answer involves some of the biggest ideas in history.

———

WHEN WE LOOK at the figure of Temperance carrying an hourglass, we are seeing the worldview of a group of medieval Italian political theorists who, in turn, had drawn their ideas from a set of Roman writers. What these writers had been considering was how human life should be lived—or rather, how civilizations should operate. And it was their belief that both temperance and time were at the heart of it.

In 44 BCE, two centuries after Rome had received its first public sundial on a column in the Forum, the Roman philosopher Marcus Tullius Cicero had written that temperate behavior—self-restraint and moderation—was exactly like time itself, neither moving too slowly nor too quickly. Cicero's contemporary Marcus Terentius Varro went further, claiming that temperance and time (*temperantia* and *tempus* in Latin) were not just like each other but were the same thing: "It is from the temperate movements of the sun and moon that time itself is named," he said.[3] Fast-forward to medieval Europe, and an anonymous twelfth-century French writer, inspired by Cicero, taught readers of the *Moralium Dogma Philosophorum* that temperance was an essential quality of timeliness.

But it was the Italian ideologist Brunetto Latini who, in the 1260s, pulled together and promoted these Roman theories into an encyclopedic work arguing that the virtue of temperance was embodied in time itself.

It is a complex set of ideas, and it was a difficult message to get across. A temperate life of moderate self-restraint cannot have seemed like an attractive option when weighed against a life lived to the full. Indeed, in the centuries before Lorenzetti sprang the symbol of the hourglass on to the Sienese scene, temperance, of all the virtues, was considered the least important. But as the fourteenth century dawned, some of the greatest thinkers in Europe had started coming around to the view that, of all the virtues that humans should live by, temperance was the most important: superior to fortitude, prudence, magnanimity and justice. Superior to hope, and to charity. Superior even, perhaps, to faith; or, rather, all these other virtues were subsumed *within* temperance. By living a temperate life, one would live by all the other virtues automatically.

But the ideas of these medieval theorists went further. In about 1334, Heinrich Suso, a monk living in today's Germany, wrote a book called the *Clock of Wisdom*, a religious work that became one of the most widely read texts in northern Europe for two centuries. In it, the character of "wisdom" from the title takes on two other guises. Wisdom speaks as temperance, and also as Jesus Christ. Suso claimed that he wanted his book to be like an alarm clock, "to waken the torpid from careless sleep to watchful virtue."[4]

And with this he had put forward an audacious proposition. Not only had temperance been elevated to become the highest of the virtues—it had been equated with Christ himself: Christ *was* temperance.

This was an extraordinary transformation of our moral code, and one that many still live by today. From being a rather drab but necessary message of moderation and self-restraint, temperance came to define, in the eyes of Europe's most influential thinkers, the most virtuous, God-like existence. To be temperate was to act like Jesus Christ.

For God-fearing medieval Europeans, this was a powerful message. And it was a mere four years after the first publication of Suso's work that Ambrogio Lorenzetti painted Temperance holding her new-fangled hourglass.

It seems as if it was Brunetto Latini's political vision of the 1260s that inspired Lorenzetti's depiction of 1338, but it was a breathtakingly new realization of the centuries-old theory. The groundwork had been put in over the years to set temperance up as the ultimate virtue and to associate it with time. The next stage was to give time a symbolic form, and that is what the hourglass provided. Lorenzetti's fresco was the first time an hourglass had been used in this way. And, by holding this new technology in her hand, Temperance was giving us a clear message. By carefully measuring and using the time that had been given to us—by being disciplined, by restraining our excesses—we would be living virtuously. The hourglass in the hand of Temperance was telling us that a temperate life was a better life.

That is what I went to Siena to see. As a historian of time, I found it deeply moving to see the oldest known depiction of an hourglass, and not just in general terms. When I had been a junior curator at London's Science Museum, I had worked with a fine historic collection of hourglasses, and one of my earliest curatorial tasks had been to carry out some research into their history. It was only a couple of days' work—I think somebody had written in with an inquiry about them, and the senior curator had given the task to me, knowing it would be a good way for me to learn about that part of the collection—but I have had an interest in hourglasses ever since. Moments like that may seem trivial, but they can leave a lasting impression. I think my work on that collection of hourglasses is one of the reasons I came to care so much about the mundane technologies of everyday life, which has informed the way I see the world and the way I write history. To see the frescoes of Siena, which are like a Genesis story for hourglasses, was a career highlight.

But what *truly* moved me was the sense that I was in a room where big themes in history had actively been played out. I was not just looking at a picture, and I was not just seeing an hourglass. Instead, I could feel something of what Heinrich Suso had intended when he wrote about his plans "to waken the torpid from careless sleep to watchful virtue." As I stood in the Palazzo Pubblico, staring at Temperance as she looked at her hourglass, I felt, somehow, like a better person. And that was because I was being worked over by a group of thirteenth-century political theorists who wanted to change the very way we live our lives.

———•———

TEMPERANCE DID NOT carry her hourglass for too long. In about 1400, only sixty years after Lorenzetti painted his Siena frescoes, the

Venice-born political thinker Christine de Pizan wrote *The Epistle of Othéa*, a lavishly illustrated treatise on statecraft and moral character that found its way into the hands of monarchs and the nobility in countries across medieval Europe, including a copy dedicated to the English king, Henry IV. It was a book about wisdom, and in its pages could be found a fabulous image of a goddess-like Temperance. But this Temperance did not carry an hourglass. Instead, she was depicted adjusting a large geared mechanical clock, topped with a bell, the wheels and pinions held within an ornate metal frame from which the clock's driving weights were suspended.

In the text that accompanied the image, de Pizan explained:

> because our human body is made up of many parts and should be regulated with reason, it may be represented as a clock in which there are several wheels and measures. And just as the clock is worth nothing unless it is regulated, so our human body does not work unless Temperance orders it.[5]

The mechanical clock was simply a better symbol for temperance. Kept wound up, these devices could run for as long as they were carefully looked after and regulated. It was this sense of regularity and the *body* that made clocks the perfect technology to preach a message of temperance, and de Pizan's book marked a turning point in the symbolism of time. From then on, the clock, not the hourglass, was used as the symbol of a temperate life, and within just fifty years had become rooted in the European imagination.

One example, out of all the candidates that could be chosen, shines out. If we want to understand the fervor of technological development in the fifteenth century, and the moral elevation of mechanism in society at that time, then we need to return to Heinrich Suso's *Clock of Wisdom*, that profoundly influential moral treatise which equated temperance with wisdom and with Jesus Christ himself. By the middle of the century, Suso had been dead for more than eighty years, but his work continued

Temperance surrounded by clockwork technology, illustrated in Heinrich Suso, *Clock of Wisdom*, c. 1450

to be republished. And an illustration in a French translation published just before 1450 truly takes the breath away.

It shows Wisdom, or Temperance, adjusting a mechanical clock, just as Christine de Pizan's goddess had done in 1400. But Suso's illustrators went so much further. Here, Temperance is pictured in a sumptuous room, decorated with fine wallpaper over an intricately patterned floor, filled with horological technology. It is like the most exquisite exhibition of timekeeping, and it would not look out of place in any museum today.

The clock that Temperance adjusts is impressive and richly detailed, with a twenty-four-hour dial and a cord leading up to the roof to operate a bell. Hanging from the clock's plinth is a large brass astrolabe, the type of complex flattened model of the night sky that had been used in civilizations around the world for a millennium by that point. Behind Temperance stands an even bigger clockwork mechanism, either an automatic alarm or a five-bell carillon, filling the room with the most captivating music. Finally, to the side of the bell machine is a table on which are displayed four types of sundial and what appears to be a portable, spring-driven domestic clock—the forerunner of the pocket watch. With such an astonishing array of temporal treasures on offer from wise Temperance, how could any mid-fifteenth-century reader resist Suso's earnest instruction to wake "from careless sleep to watchful virtue"? These clocks made people better. They made a better civilization.

———·———

NOW THAT THE hourglass had been released from the hands of Temperance, its characteristics of brevity and finitude made it the perfect candidate to take on a new role. Rather than representing a timely virtue, the hourglass started to become a symbol of the passage of time itself.

In his fourteenth-century work, the "Triumph of Time," the Italian poet Petrarch invited us to reflect on the insignificance of human life:

> *What more is this our life than a single day,*
> *Cloudy and cold and short and filled with grief,*
> *That hath no value, fair though it may seem?*
> *. . . And fleeing thus, it turns the world around*
> *Nor ever rests or stays nor turns again*
> *Till it has made you nought but little dust.*[6]

Petrarch was cautioning against the perils of fame and vanity, reminding readers that nobody—not even the rich or famous—knew when their life would end, and that we would all end up as just a little dust.

The hourglass, with its lower vessel catching the sand of time as it ran out to leave only dust, was heaven-sent to symbolize Petrarch's message. Petrarch himself did not give us a personification of Time, but the fifteenth century saw his work copiously illustrated and re-illustrated, and soon his vivid words had been given visual form by illustrators in the flesh-and-blood depiction of a new character for European culture.

We know it well today. It is the figure of Time as an old man—Father Time—with wings and a beard, accompanied by an hourglass.

For hundreds of years following its introduction by the fifteenth-century illustrators of the "Triumph of Time," the symbol of Time with his hourglass permeated Western art. Wherever an allegory of time's destructive effects was needed, there was a winged bearded man close by. The hourglass was the perfect symbol. At a glance, it showed a life span: how much time had already passed, and how much was left to come. It depicted the inexorable finitude of life. So it is no surprise that soon after it appeared with Time in the "Triumph," the hourglass took on an additional symbolic role, and this one was the symbol to end them all: that of Death itself.

———

THE YEAR IS 1776, and a horrific, nightmarish scene unfolds in the great cathedral of Notre-Dame de Paris. On a high plinth is the tomb of Henri-Claude d'Harcourt, a lieutenant-general in the French army who had died seven years earlier. In front of it are the trappings of his distinguished career: his sword, shield and helmet; the creased fabric folds of a military banner. To one side is the winged figure of a boy carrying a torch, a guardian angel watching over his master. Standing in front of the scene, looking up at the raised coffin with her hands clasped in joy and an expression of growing ecstasy on her face, is Marie-Magdeleine Thibert des Martrais, d'Harcourt's widow.

And there, as she looks on, is the grisly sight of her late husband, emaciated by death, attempting to clamber out of his tomb. But this is not the most shocking sight of all. For behind him, standing ominously over the struggling cadaver, its face a rictus of horror, is the shrouded skeleton of Death, the grimmest of reapers. And clutched tightly in the bones of Death's right hand is the message that tells Marie-Magdeleine what she so desperately wanted to hear: that the final hour has come, and that husband and wife can be reunited. What Death brandishes is a simple hourglass.

This scene, sculpted with extraordinary skill in white marble by the French sculptor Jean-Baptiste Pigalle, had been commissioned by d'Harcourt's grieving widow as a monument to marital fidelity, and it took five years to complete. By 1776, when the tomb was finally revealed, the skeletal figure of Death had already been stalking with an hourglass in his hand for 300 years since the two had become intertwined in the artistic imagination in the late fifteenth century. The hourglass, once an encouraging symbol of temperance, had become an icon of human mortality.

By then, Death, holding his hourglass proudly and defiantly aloft, could be seen everywhere. D'Harcourt's tomb sculpture at Notre-Dame is a particularly fine example, but it was not just the high-born who memorialized their dead with symbolic reminders of the need to live virtuous lives that could end at any moment. From the sixteenth century until the eighteenth, not only tomb sculptures but countless simple gravestones across Europe were incised with the image of the hourglass alongside other symbols of death such as skulls, coffins and bells. Visit windswept Scottish and Irish graveyards today and you will probably find an hourglass.

It did not act only as a symbol of virtue. On the high seas of the eighteenth century, piracy was rife. We think traditionally of the skull and crossbones as the symbol on the flag of a pirate ship, but the hourglass accompanied many of these brutal voyages, too. One 1724 book chronicling the history of piracy described how the flag of a particularly notorious pirate ship depicted "a Death in it, with an Hour-Glass in one Hand, and cross Bones in the other, a Dart by it, and underneath a Heart dropping three Drops of Blood."[7] Imagine the sense of terror that would strike the crew of a merchant ship as the macabre symbol came into view.

But it was in art that the figure of Death with an hourglass made its presence most keenly felt. A type of allegorical scene often known as the Dance of Death or *Danse Macabre* emerged that reminded viewers that no matter who they were, or how high their place in society, death would come for them in the end. It was Hans Holbein's extraordinary depictions of the Dance of Death, made in about 1525 and published in 1538, that pressed the hourglass firmly into the bony hand of a skeletal, ever-

The Triumph of Death, painted by Pieter Bruegel the Elder, c. 1562

present, grinning, almost gleeful Death. Then, about twenty-five years after the publication of Holbein's images, and knowing them well, Pieter Bruegel the Elder painted a scene that must rank as one of the most hellish in the history of art.

The Triumph of Death, a large, brightly colored oil painting made in the 1560s, now in the collections of Madrid's Museo del Prado, is a truly horrific scene. It shows an army of *Danse Macabre* skeletons that have risen from the grave and are roaming over the landscape, torturing and slaughtering everyone in their way. There is no escape from the bloodbath being meted out in Bruegel's sickening scene. Skeletons hang some of their victims from gallows and behead others. Skeletons slit people's throats, hack their way through crowds, stab and burn people alive. Skeletons force people to eat the flesh of those that have already been killed. One skeleton wears the face of a victim, flayed from its skull, its eye sockets hauntingly empty. Skeletons use their coffin lids as shields and ride rotting horses over people attempting to flee the carnage. A pile of dead and dying victims grows at the center of the hideous scene, and bloated corpses float in a nearby pond as more bodies are tipped in to join them. A thick, dark-red slick of blood spreads across the water from a decapitated head that has been dropped in.

At the left of the scene are two eye-catching skeletons. One sits behind a fallen emperor, still clad in his armor and fine robes but clearly badly injured. Perhaps the skeleton is dragging him backward. The other note-worthy skeleton in the front of the painting rides slowly into the center of the view, its emaciated horse pulling a wagon filled to the brim with human skulls, its wheels crushing and shattering the limbs and torsos of the unfortunate victims trapped underneath. In the hands of each of these gleeful, murderous, vile skeletons is an hourglass.

Death, carrying its hourglass, does not *stalk* this landscape, it *hordes across* it. Old and young, rich and poor, men and women, religious and secular, people of all stations and from across the land perish, their time cruelly and violently ended by the mocking skeletal figures of Death.

———

THERE WAS LITTLE escape from Death and the hourglass as the centu-ries passed. Still-life paintings, known as vanitas pictures, hung on the walls of sixteenth- and seventeenth-century homes in the Netherlands and elsewhere. Beautiful earthly treasures, from fine clothes and fur-nishings to jewelry and objets d'art, were commonly depicted alongside skulls, rotten fruit, snuffed-out candles and, of course, hourglasses. But one anonymous sculptor chose to represent the theme in a more visceral way. In the collections of the Science Museum, in London, is a human head modeled, life-size, in wax. Dating probably from the eighteenth century, it can be viewed from the sitter's left side or right.

Her left side shows the face of a beautiful woman, her complexion a clear milky-white, her hair softly curling, a graceful hand carrying a posy of freshly picked flowers draped on her shoulder from behind. Viewed from the right, though, her skin has rotted away, revealing a hideously grinning skull. A giant worm crawls from her eye socket and a large insect skitters away. Maggots infest the rotting remains of her hair. A frog and a snail sit by her neck, and it is now the skeletal hand of death that clutches her from behind. There is no hourglass in this scene—there would scarcely be room alongside the piles of flesh-eating creatures—but on a plaque at the front of the sculpture, crudely written, are words from Ecclesiastes 1:2, which translate as follows: "Vanity of vanities, all is vanity."

Markets

Stock Exchange Clock, Amsterdam, 1611

Ömer Ağa stood in the middle of Amsterdam's Dam Square surrounded by his nineteen-strong party of advisers, interpreters and hosts, and gazed toward the huge new trading exchange that straddled the mighty Rokin canal, just to the south of the square. The year was 1614, and Ağa was on a fact-finding mission to the Dutch Republic as the Ottoman Empire's newest diplomatic emissary. Top of his list of must-see sights was this bold new building, completed just three years earlier. It was hard to miss, as it was the size of a soccer field and could accommodate thousands of traders in its 200-by-115-foot enclosed inner courtyard, but what Ağa really noticed was the four-sided clock tower that loomed over the vast structure and the streets and canals all around, as well as the booming sound of its bells when they rang out the hours and then, at noon, tolled repeatedly for a few minutes before falling silent. Little did they realize it, but Ömer Ağa and his retinue were listening to one of the most significant clocks ever made. It was fitted to the world's first stock exchange, and it was sounding the birth of modern capitalism.

Amsterdam Stock Exchange, engraved in 1612

FROM THE MOMENT the Amsterdam exchange building first opened its doors in August 1611, traders were forbidden from trading anywhere else in the city. But the exchange did not just put spatial boundaries on trade. It concentrated traders in time, too. A few days before the new facility opened, the city council had issued a bylaw proclaiming that trading could only take place between the hours of 11 a.m. and noon, Monday to Saturday. At noon, the clock installed in the tower high above the exchange building would toll a bell for seven and a half minutes. If any traders were still in the exchange, or in the streets nearby, they would be fined. Additionally, trading was allowed between 6:30 p.m. and 7:30 p.m. during the summer months between May and August, and in winter evening trading took place for a thirty-minute period marked by a tolling bell at the city's gates. At the end of evening trading, the exchange clock would again sound for seven and a half minutes and fines were issued for anyone caught trading after the bells fell silent.

Why were such strict limits placed on trading at the Amsterdam exchange? There were several reasons. One was a practical problem familiar to anybody involved with trade in a busy city center: time limits reduced congestion and disruption in the streets nearby. Another was that clocks made trading more efficient. Short, fixed trading hours concentrated buyers and sellers together, making it easier for each to find enough of the other. This increased the volume of trade, which was good for traders and for the city council collecting taxes on transactions. But clocks also helped prices to remain fair, as they could be used to regulate the people who occupied intermediate roles in the functioning of a market.

Some of the earliest references to mechanical clocks being used in towns and cities, in the Middle Ages and soon after, related to market restrictions. The first urban markets brought producers of food, cloth and so on into direct contact with the consumers of their wares. But as towns and cities grew, this model started to break down. It stopped making sense for every producer in the countryside to make the journey all the way to the center of towns. So, "intermediate trading" emerged, whereby third parties might buy up the goods from several small producers somewhere on the edge of town, before bringing them in and selling them themselves at the market. Soon, a whole range of intermediate roles sprang up. Wholesalers, merchants, shopkeepers and peddlers were some, but intermediates also included financiers who advanced funds, and those speculating on the future in the hope of offsetting risk (whether because of bad harvests or other unpredictable events) and making more money. Some people occupied more than one role.

As populations grew and moved in increasing numbers to towns and cities, and markets began to sell more and more products, the rise of intermediate roles in market-based trade was inexorable, creating a new stratum of people who neither produced goods nor consumed them, but traded, speculated, brokered, hoarded, flipped and financed. Some market authorities feared intermediates would drive up prices or limit supplies and turned to clocks to control their involvement. Clocks meant that different groups could be treated differently at the market. In a sixteenth-century grain market, for instance, the first hours of trade could be restricted to residents, before bakers of bread could get in, and then the pastry bakers could enter. Only after several hours were wholesalers and other intermediate traders allowed in. But as societies and their market trading became ever more complex, the role of intermediates like brokers and financiers became increasingly important in keeping the flow of trading running smoothly. And, before long, finance became something that could be traded in its own right, and clocks took on a new regulatory role.

Amsterdam's was not the first trading exchange. Antwerp and London had had exchanges since the sixteenth century where goods and money were traded, but Amsterdam was the first of a new kind of exchange: what became the modern securities exchange. As well as being a place to trade in commodities like salt or hides, people could also buy and sell financial assets. It started out as a place to buy and sell shares in the Dutch East India Company, an early joint-stock company and the first with freely tradable shares, but soon was used to trade other company shares, futures contracts and insurance policies as well as becoming the place to go for information about the state of the markets. The financial market had arrived, but its products, and the prices paid for them, were time-dependent. The time at which each securities transaction was made, or would be enacted in the future, was central to this new type of trading; to work fairly, everybody had to agree what time this was. In other words, trading needed time stamps, which is where the exchange clock came into its own. Clocks were no longer about excluding intermediates from the market. In the new exchanges, intermediates *were* the market—with the clock watching carefully over the whole thing.

I VISITED AMSTERDAM in 2020 with the same group of clock friends that had taken me to Chioggia in 2018 and Siena in 2019. This time, we stood in Dam Square and looked toward the old Rokin, just as the Ottoman envoy Ömer Ağa had done when he visited with his diplomatic party four centuries earlier. The original exchange building is long gone. It was extended in 1668 and its clock tower taken down, and the building was demolished entirely in 1836 because it had subsided and become too expensive to maintain. You cannot even see the water of the Rokin canal that once flowed under the exchange, as the channel was filled in during the 1930s and is now covered by a shopping mall. Nevertheless, my friends and I explored the streets and buildings around the site of the former exchange and scrutinized old maps and vistas of the city so that we could conjure the building back to life in our minds. We imagined the clock tower looming over us, the booming sound of its great bell, and the shouts and bustle of traders entering and leaving the exchange, their fortunes made a little greater or less from their day's trading.

But we had an additional reason for visiting. We had heard rumors that the original clock from 1611—the clock that kick-started modern capitalism and kept the world's first stock exchange on time for the first half century of its existence—might have survived. In 1668, when the building was extended, a new clock had been installed at the exchange, and the old mechanism was sent a mile eastward to be installed in the city's new Oosterkerk, or Eastern Church, which was just about to be built and which opened three years later. And that is where my friends and I walked next.

We were met at the church by the Dutch clock expert Rob Memel and the church's director, Henk Verhoef, who led us up flight after flight of ever steeper staircases and ladders until we reached the very top of the building's tower, where we stepped into a huge loft. And there, sitting in the gloom over in a far corner, were the remains of the Amsterdam stock-exchange clock.

It is hard to imagine that Ömer Ağa in 1614 was any less impressed by this clock than we were in 2020. This historic mechanism stands six feet high on a massive wooden trestle. Its wheels, pinions and levers are held within huge pillars made from the finest wrought iron, imported from Sweden, the best producer in the world at that time. It was clearly made to impress visitors by its scale and the quality of its design and construction, far better than comparable clocks then being made in Britain. This was the state of the horological art, just right for a building that would change the modern world. The clock is no longer connected to the Oosterkerk's dials, nor to the church's huge seventeenth-century bell that can still be heard across Amsterdam. It has been replaced by a more modern clock mechanism, and some components of the original have been lost along the way. But the part that originally struck the Amsterdam stock-exchange bell every day from 1611 to 1668, calling the traders together and regulating their business, remains in this lofty church tower, a technical triumph that holds within its old wheels and pinions the pioneering spirit of capitalism.

WHEN WE TALK today about the financial markets, it is usually an abstract idea. We no longer mean real, physical places where people bargain face to face. Instead, we mean the financial system in which billions of transactions take place, day in, day out, dispersed over the globe and barely comprehensible as a *thing*. The market economy is the basis for the global capitalist system. But this unseen entity emerged from real, physical markets, regulated by time as dictated by clocks. It would be easy to conclude that real clocks are no longer needed now that the markets are formed of global computer and communications networks, with buy and sell orders for everything from grain and oil to exotic financial instruments flashing around the world on fiber-optic cables at the speed of light. Easy, but wrong. The markets, and their clocks, are just as real, though dispersed, faster and *much* more complex.

If markets are the key mechanisms of global capitalism, clocks are the key machines of markets. Capitalism is an arms race of timing, and clocks make markets work now more than ever before. But there has been a long journey to get where we are today. Clocks have always loomed large over the halls of finance. Clocks on high towers, clocks high up on the walls of banking and trading floors, clocks ticking and tolling their bells, always in view and in earshot. Clocks stamp the transactions and call the traders together.

London's first exchange building, the Royal Exchange, was built in the 1560s. Most of its traders sold goods but, at least for a while, before they were evicted for rowdy behavior, the financial traders met there too, and later it became the home for trading insurance policies, and through its history we can see how timekeeping technology kept pace with the expansion of the financial marketplace. The original Royal Exchange incorporated a tall tower that rose high above the streets below, crowned with a huge grasshopper, the emblem of Thomas Gresham, the merchant who financed the Exchange. Inside the tower was a clock, which tolled a bell at noon and 6 p.m. to gather the traders together. According to reports a few years later, the clock was always out of order, partly because the building had been hurried through and was not well built. In fact, the whole Exchange perished in the Great Fire of 1666.

When it was rebuilt afterward, the true importance of timing for the efficient operation of markets was better understood. The new clock was set up under the supervision of Robert Hooke, one of the seventeenth century's leading scientists and a pioneer in precision horology. It was much better than the first, and played a different tune, four times per day, for every day of the week. Unfortunately, in 1838, this building too succumbed to fire. As the flames licked ever higher, Hooke's clock just finished chiming the popular eighteenth-century Scottish tune "There's Nae Luck aboot the Hoose" before the peal of eight bells crashed through

London Stock Exchange trading floor and clock
c. 1800, engraved c. 1878

the structure, demolishing the building's entrance arch. By then, the
building was home to Lloyd's of London, insurance underwriters. A lot
of money was lost that night. A new exchange was completed in 1844
and this time its clock was designed by the Astronomer Royal and kept
time to within a second. Before Big Ben began sounding across London
in 1859, the Royal Exchange clock was considered one of the best public
timekeepers in existence. It played twenty-one different tunes and it was
impossible to ignore—it could be heard for miles around.

It is hard to find a historical depiction of a stock exchange or bank—

painting, engraved print or photograph—that does not include a clock prominently in view. But the nature of these clocks changed through time, as other technologies speeded up the trading: electric telegraphs, steamships, computers. The faster the trading, the better the time-stamping needed to keep track of the trades.

In 1886, the Standard Time Company, a London firm that sold accurate time signals over electric telegraph wires, published a list of its subscribers. These were people whose businesses relied on knowing the right time, and every financial institution in the city was included. Electric time signals arrived at the London Stock Exchange, the Bankers' Clearing House, Lloyd's of London, the Wool Exchange, the Baltic Exchange, the London Corn Trade Association, the National Life Office, the Northern Assurance Company and twelve banking headquarters, including today's HSBC, RBS, NatWest, Barclays and Lloyds. Each subscriber operated a network of clocks—numbering in the hundreds for large buildings with many rooms—all interconnected by electric wires.

The London Stock Exchange was the perfect example, and one of its automatically corrected clocks has survived in a private clock museum in London. It is an oversized and finely constructed mechanical clock made in the early nineteenth century by the noted clockmakers Thwaites and Reed, so it was already decades old by the time it was fitted with a Standard Time Company synchronizing device in the 1880s. From then on, each hour, an electrical signal was received from the time company which flashed across a similar device fitted to every clock on the circuit, causing it automatically to correct the clock to the right time. It was accurate to within about a second of Greenwich Mean Time, the official UK timescale, according to experiments carried out on surviving equipment.

The Standard Time Company's clock-synchronizing service was not cheap to subscribe to, and the maintenance of this high-tech network was onerous—it was fragile and vulnerable. Yet hundreds of companies, large and small, went to the trouble to get it, and many were in the financial sector. Accurate time was increasingly valuable, and businesses would seek out the latest horological technology that would help them in their ambition to make money.

By the time the Standard Time Company began synchronizing clocks across the City of London, financial markets were no longer the single locations fixed in time and place that the Amsterdam and other early exchanges had been. They were geographically and (therefore) temporally dispersed. Trading on the markets meant engaging with people, institutions and technologies worldwide.

When the concept of daylight saving (or shifting clock-time forward an hour during summer months) was first put forward in the early years of the twentieth century, both the London Stock Exchange and the Liverpool Cotton Market—Britain's first futures market—were worried about the proposed scheme. It came down to the temporal nature of markets. The UK markets both closed at 4 p.m., partly out of convention but, in the case of London, to catch the evening postal collection which processed the thousands of letters and memoranda that formed the record of its daily trading.

A vast amount of that daily trade was with the New York Stock Exchange, which closed at 3 p.m. Owing to the five-hour time difference between Britain and New York, that meant there was just one hour each day when both territories were open together: 3 p.m. to 4 p.m., UK time, which was 10 a.m. to 11 a.m. in New York. The daily trade during that single hour was, in the words of one broker during the discussion, "immense."[1] If Britain's clocks shifted forward by one hour in summer, the window would close completely, and Britain's brokers feared that trade would drift to another global exchange better suited to fit with New York's hours. Of course, London could stay open later, but it would miss the post each evening, making it less efficient, so trade would drift away, and neither exchange could *open* an hour later, as they were tied into other temporal windows around the world at that time of day, so working hours would have to increase, so clerks and traders would drift off to other employments, and so on. Alternatively, New York could have changed its hours but it too was tied into an infrastructure that would start to unravel.

When the Liverpool Cotton Exchange moved into brand-new permanent premises in 1907, brokers were surrounded by a network of synchronized electric clocks in the trading hall and surrounding offices,
designed, so the Liverpool press explained, "in order to obviate confu-
sion and conflicts of opinion in the transaction of business."[2] A little
over a century later, synchronized networks of clocks have moved on,
somewhat. When Benjamin Franklin, in 1748, said that "time is money,"
he could not have imagined how his maxim to work hard and be thrifty
would take on new meaning centuries later.[3] Today's financial markets,
which in total trade hundreds of billions of dollars per day, need clocks
that are accurate to 100 millionths of a second.

———·———

BROADLY SPEAKING, THERE are three ways to buy or sell financial
instruments (like stocks or futures) today. The first is human trades,
where a person—a real human—issues an order to buy or sell a certain
amount of a certain instrument online or over the phone. It is a human
finger on the mouse button or a human voice shouting orders on the
phone and, apart from the technology, it is the same way things were
done in 1611 at the Amsterdam exchange. You issue your instructions and
your order gets in line with everyone else's. It is nice to think this still
happens today. Think of it like heritage trading.

The second way to trade on the financial markets is to use computers
running complex sets of instructions, or algorithms. These algorithms,
or "algos," automatically issue buy and sell orders according to pre-
programmed rules. If the price of this stock comes down below this level,
then buy this much, for instance. If this company reports a loss in its
quarterly accounts, then sell this percentage of our holdings. That type
of thing. Algo trading, which started taking off in the 1990s, is now big
business because it happens faster than humans can react, and, remember, this is an arms race of timing.

But conventional algo trading runs at a snail's pace compared with the third method of buying and selling on the financial markets. A subset of algo trading techniques known as high-frequency trading, or HFT, entered the scene in the 2000s and now accounts for over 50 percent of stock trading in the USA—a little lower in the UK. As its name suggests, HFT is all about speed, making huge numbers of trades every second. The individual trades are usually very small and the profit on each is also tiny—but because of the sheer volume of them, the profits can build up into a sizable amount. At the end of each trading day, HFT traders aim to hold no assets. They sell everything they buy that day. This is not about building up a stock portfolio or holding a futures contract for weeks or months, it is about the friction of trading itself generating money. The assets being bought and sold are of no interest to the HFT traders. They could be company shares, domestic mortgages or exotic financial derivatives. The point is that the computers spot the chance to make a tiny profit—a chance that might last only the tiniest fraction of a second—and they pounce, at light speed.

Where do clocks come into this? The first way is in the technical synchronization of the computer networks over which the trading takes place, which, as we will see in a later chapter, relies heavily on accurate clocks to work properly. But the second way in which clocks have become important in the era of HFT and algo trading is in many senses the same as in 1611 in Amsterdam. There, the clock was used not just to bring buyers and sellers together but also to time-stamp financial transactions to ensure everything was done officially and above board. Physical markets with clocks allowed oversight by regulators—in that case, the city council. Anyone who broke the rules would be fined. It is the same in today's financial markets. Clocks enable regulation to happen.

The risk today is that firms with computers trading at almost light speed might be able to "see" the financial markets a tiny split second earlier than their competitors, which would enable them to profit on the information they saw in that tiny window of time. In some circumstances this would be illegal, so what regulators need is to be able to read the sequence of every trade and every data feed, to make sure that nobody is breaking any rules and unfairly jumping the queue. Assuming everyone in the marketplace agrees on what time it is, then the trades and data feeds can be time-stamped, and it is clear what happened when—and what came before or after what. It means that regulators can keep a watchful eye on the conduct of the markets.

It sounds easy but, in practice, there are three problems. The first is agreeing on a timescale. In today's world many timescales are in use. We have TAI, which comes from atomic clocks, and UT1, which comes from the Earth's rotation, as well as UTC, which is a hybrid of the two, but there are seventy-five very slightly different versions of that depending on where you are in the world. Or we could choose GPS time, or time from other satellite navigation systems such as GLONASS, BeiDou and Galileo. We have, too, time from radio stations, internet time services, cell-phone providers and broadcasters. Deciding which timescale is "the" time is not a trivial problem to solve.

The second problem with time-stamping in the financial markets is a hardware problem. Once you have accepted which timescale is the one to use, how does every market participant get access to that timescale so that all the time stamps agree with each other? It is easy enough to get the time to a clock on the wall of a stock exchange for those human traders, but what about the thousands of computer servers exchanging high-frequency trades around the world? Each server, each microprocessor, each network switch needs to have access to the same clock, somehow. Again, this is not trivial, to say the least.

The third problem is in some ways the most challenging. Imagine that two financial orders were placed half a second apart. If the time-stamp clock ticked once per second, as many clocks do, it would be quite possible that both trades would hold exactly the same time stamp, and nobody would know (or be able afterward to find out) which came in first, opening the system up to abuse. The third timing problem therefore is the precision of the time-stamping clock—how small its timing intervals are—as well as its accuracy. The more precise and accurate the time stamp, the more likely it is to get the trades in their proper order.

But of course more precision and accuracy is harder to achieve and even harder to distribute over the whole network.

It is important to be clear about what is needed here. The European directive on markets in financial instruments which came into force in January 2018, known as MiFID II, selected UTC as its timescale, and demands time stamps precise to one second for human trades, or one-thousandth of a second for normal computer algo trades. But for HFT trades the time stamp must not deviate from true UTC by more than 100 microseconds, or millionths of a second, and it needs to have a precision—the gap between two successive stamps—of no more than one-millionth of a second. Put another way, instead of stamping once per second, MiFID II–compliant clocks stamp a million times each second. And those time stamps need to be found on every chip in every computer server in every trading exchange across the entire European financial market.

Just think about that for a moment. All the computers that are involved with financial trading across the whole of Europe need to show the same time as each other to *100 millionths of a second*. There are thousands upon thousands of such computers in a single data center alone. Records need to be kept documenting the time stamps at every moment, in every location throughout the network, for future audit. The requirement is for at least five years; the best time suppliers keep them for seven. There are 221 million seconds in seven years. A million time stamps per second. This is a requirement for the most astonishingly accurate, precise and reliable time, twenty-four hours a day and documented for several years in case the regulators want to check something, and any firm that does not comply risks being slapped with a fine of 10 percent of its global revenue. Not profit—revenue.

———————

THE AMSTERDAM STOCK Exchange clock made in 1611 would have been accurate to better than half an hour a day. It kept time using a horizontal bar or wheel that rotated first one way and then the other, powered by falling weights and regularly corrected by its keeper using a sundial. In the 1650s, a new horological technology entered the scene, when pendulums were first used to control clock timekeeping. They were a radical improvement in accuracy because they oscillated at a frequency that was predictable, unlike the older clocks, and the clock-making industry jumped on the chance to develop and refine them. By the 1920s, the very best mechanical pendulum clocks had reached an accuracy whereby they would gain or lose no more than a second in two or three months compared with true time as measured by the astronomers at national time-finding observatories.

Then, electronics engineers started building quartz clocks. These used pieces of quartz crystal that vibrated like a bell, and the frequency of the vibrations could be measured with electronic equipment, so they could be used as a clock. These were much more accurate than pendulum clocks—good to about a second in thirty years to start with and soon much improved.

But things really started to change after 1955, when the UK's National Physical Laboratory, or NPL, built the world's first successful atomic clock, which used fundamental and unchangeable properties of atoms for its time base. No longer did the world's best clocks take their time from the shape and size of a physical oscillator such as a pendulum or a slice of crystal, which put practical limits on the quality of timekeeping. With atomic clocks, the accuracy and precision that could be measured was almost limitless. And since that pioneering NPL clock first began to keep time in 1955, physicists have provided the world with a procession of exponentially more accurate clocks.

The 1955 atomic clock was accurate to one second in 300 years. By the 1980s, NPL's atomic clocks were keeping time to an accuracy of one second in 300,000 years. *Homo sapiens*, the human race, appeared about 300,000 years ago. It is a long time to keep time within a second, but today's atomic clocks are much more accurate than that. Clocks now keeping time at NPL, known as cesium fountains, are good to one second in 158 million years. But even these timekeepers are sluggish compared with the next generation of atomic clocks the scientists and technicians are now developing. These have an accuracy of plus or minus one second in 30 *billion* years. That is more than twice the age of the universe. Or, to put it another way, if one of these clocks had somehow been set running at the Big Bang—the explosion that brought the entire universe, even time itself, into existence—it would currently be wrong by less than half a second.

But how do we set *our* clocks right? NPL does not just keep time. It gives it away: NPL clocks feed into radio time signals and internet time services that allow all of us to know the time they keep. But neither of these is accurate, precise or stable enough for MiFID II, with its demand for time stamps every millionth of a second. The clocks at Teddington are much more accurate than that, but the hardware that transmits the time to the outside world could not cope. Time signals from GPS satellites would be one alternative, and some financial companies use GPS time for their stamps, but the signals are technically vulnerable. As MiFID II was published and the financial sector scrambled to build systems to comply with it, NPL spotted a gap in the market. Leon Lobo, an NPL business development manager, recently said, "At the microsecond level, I would say no-one in the City of London has the same time."[4] His job is to sell it to them.

———————

THE FIRST THING that hits you is the noise. The rooms full of six-foot-tall, black-painted steel-mesh cabinets filled with computer servers do not hum, they roar. They are surrounded by fans and air-conditioning equipment to keep the equipment cool and the noise is so great you must shout your conversation, not that most people visiting these facilities are there for a chat. Telehouse North, which opened in 1990 near the old East India Dock in London's Tower Hamlets, was Europe's first purpose-built data center allowing businesses like those at nearby Canary Wharf to site their computer servers in the same building as those of their customers, partners, exchanges and super-fast connections to other facilities around the world. Now, the area has become a bleak city district

filled with gigantic windowless structures, bristling with razor-wire
fences and CCTV cameras.

Below the streets that divide up these structures is a dense network
of fiber-optic cables over which the light-speed commands of modern
finance pulse, day and night, serving the financial powerhouses nearby.
But there is a single pair of fiber cables running into Telehouse North
that is worth looking at more closely, as it has come a lot further than
Canary Wharf. In fact, it goes all the way to NPL at Teddington, fifteen
miles to the southwest.

I was at Telehouse North with an NPL scientist, Ali Ashkhasi, one of
a small team running its fiber-optic time service; I wanted to see what
the modern equivalent of Amsterdam's 1611 Stock Exchange clock looked
like. It is easy to assume that modern-day time signals, like cloud com-
puting or electronic financial trading, are somehow virtual; that this all
exists in what we call cyberspace, as if it has no material form. The reality is very different. This is a world of big, heavy, loud machines, using
vast amounts of electricity and requiring industrial-scale cooling, which
occupy huge buildings covering acres of valuable city real estate. It is a
world of steel and concrete, bricks and mortar, security guards and staff

Telehouse data-center complex, London, photographed in 2020

canteens. As I discovered only too clearly during my visit to Telehouse with Ashkhasi, a lot of his working day is spent filling in forms, signing out passes and keys, and finding parking spaces. The lifts smell of lasagna. Cyberspace and the services that use it are decidedly real, not virtual. And so are the clocks that regulate today's financial markets.

The single pair of fiber-optic cables runs from NPL in Teddington on a circuitous forty-six-mile route through central London out to Telehouse North. First set up in 2014, the cables carry nothing else except time. At the Teddington end, two hydrogen maser clocks, each the size of a small domestic refrigerator, provide the time to a further pair of clocks known as the Grand Masters (one primary, one backup). The masers in turn are part of NPL's ensemble of super-accurate atomic clocks that together form the UK's version of UTC (which is checked against every other country's UTC once a month in a laboratory near Paris in France). The Grand Masters feed their time into the fiber link. At the Docklands end, the time signal feeds into an identical pair of clocks in NPL's steel cabinet buried in the Telehouse data center, which become, in their turn, the Grand Masters for all the customers subscribing to the time service.

Each subscribing business has its own clock that is set right by the Grand Master, and that is what makes the time stamps for its own trading network. These are all real clocks. The equipment is manufactured by companies like Microchip and Meinberg, hardly household names but perhaps we should become more familiar with them given how much they rule our lives. There are a lot of these clocks, and they are all around us—literally, in anonymous buildings and control boxes throughout our landscape—but we never see them, unlike the tower clocks in the first stock exchanges such as in Amsterdam. There is a huge amount of money resting on them, because they make markets work, and capitalism is built on markets.

Time stamps to a millionth of a second, complying with MiFID II, are already outdated. Ali Ashkhasi at NPL told me his time system would be able to synchronize the clocks across a computer network to 100 *billionths* of a second if required, because some firms are now trading on the financial markets at the speed of a nanosecond.

5

Knowledge

Samrat Yantra, Jaipur, 1732–35

The clattering, hammering and shouted commands that accompany any large-scale building project were ever-present across the Indian city of Jaipur in 1732, as the great new state capital continued to take shape and grow. The foundation ceremony for the city had taken place five years previously, and since then progress had been rapid. Already, this gigantic construction project had transformed the former floodplain beneath the hills of Amer into a vibrant urban settlement. But the clamor of construction was muted in the courtyard alongside the city's vast new palace complex. This five-acre open space was surrounded by low walls protecting it from the bustle in the streets outside, creating an oasis of tranquillity. It was the ideal place to think.

Today was an important milestone in the urban development project of Jaipur. Maharaja Sawai Jai Singh II, ruler of the kingdom of Jaipur, stood in the quiet courtyard with his trusted construction manager and city planner, Vidyādhar, who had been tasked with marshaling the enormous resources of labor, materials and expertise needed to build the new capital city from scratch. Jai Singh's 120-acre palace development, at the geographic and cultural center of the new city, was well under way. But the king's lavish new apartments were not the primary focus of Vidyādhar's attention that day. He was there to discuss the next phase of construction work on the Hindu ruler's astronomical observatory that occupied the walled courtyard site in which the two men were standing. The calm of the serene and protected enclosure was about to be broken.

In the king's hands was an intricately carved wax model. Handing it to Vidyādhar, he told the planner to show it to the builders so they could grasp the enormous scale of the challenge facing them. What Jai Singh required, and had modeled in the wax miniature, was a huge edifice taking the form of a great triangular wall, to be fashioned out of stone and plaster, aligned perfectly north to south, with its upper surface angled exactly to the latitude of Jaipur so that it was parallel with the Earth's axis. It was to occupy a footprint of 130 by 144 feet, dug more than nine feet into the ground, and to rise seventy-five feet into the air (one-quarter of the height of London's Big Ben tower, the construction of which began a little over a century later). He required it to be topped with a roofed pavilion in which an observer could work, protected from the hot Jaipur sun. A long flight of steps running up the center of the triangular wall toward its summit was specified, allowing astronomers, including Jai Singh himself, to make careful observations. And alongside this monumental central angled "gnomon," or shadow indicator, were to sit two lower structures known as quadrants, onto whose curved surfaces the shadow of the gnomon fell, with finely marked scales divided into two-second intervals. The real challenge was to align it perfectly, to tolerances of just a few millimeters, so that it could keep time to within a couple of seconds. It took Vidyādhar and his team of builders three years to construct Jai Singh's Samrat Yantra, or "supreme instrument," and it is still, to this day, the world's largest sundial.

Samrat Yantra sundial at Jai Singh's Jaipur observatory, photographed in 1915

———————

CONSTRUCTION OF THE observatory complex at Jaipur had begun in the 1720s, and by the end of the decade over half a dozen huge masonry instruments had been erected there, some for finding time, others for observing stars and other heavenly bodies. But none were to match the scale and cultural impact of the towering sundial centerpiece that Jai Singh had always planned for the facility. When the Samrat Yantra was completed in 1735, it was impossible to ignore.

A few miles to the north of Jaipur was the old city of Amer, a place from which traders traveled to the industrial center of Sanganer, to the south of the new capital, before returning laden with goods. At the same time, pilgrims from the city of Agra, 140 miles east of Jaipur, would make the long journey toward the sacred sites of Ajmer in the west along a route that passed just over half a mile from the walls of the observatory complex. The bold new structures of the emerging capital city were clearly in view to all these travelers. The city's residents, too, were not the usual mix. Most were involved with trade and banking, helping fulfill Jai Singh's ambitions for his new capital to be a financial powerhouse.

For these influential visitors, merchants, pilgrims and residents alike, the trappings of wealth and power exhibited in Jai Singh's new palace would have been impressive, but not unexpected. However, a five-acre astronomical observatory, with a monumental sundial at its heart, was quite a different matter. One mid-eighteenth-century visitor described the Samrat Yantra as "astonishing," commenting that on its summit "there is an observation tower that overlooks the whole town and so tall that one cannot be there without one's head turning."[1]

It was not Jai Singh's first observatory. In 1721, before embarking on construction of the new capital city of Jaipur, he had built an observing facility seven or eight miles outside Delhi, where he was then based. Later, he built smaller versions of his stone observatories: at Mathura, ninety miles south of Delhi; Varanasi, 425 miles to the southeast; and Ujjain, 380 miles to the south. Why was astronomy so important to Jai Singh? Many rulers, when they took power, wanted to mark their place in history—and prove the legitimacy of their sovereignty—by getting calendars and astronomical tables drawn up based on the dates of their reigns. It showed that they had arrived and that they were in control. But what most rulers did was to revise existing tables, often dating back centuries. Only the most ambitious would establish five monumental state observatories to re-make the measurements from scratch. And Jai Singh was nothing if not ambitious. He wanted to show the world that he was at the center of the universe.

———•———

ASTRONOMY IS AN international science. Look at the world's most significant astronomical observatories today. They bring together researchers from around the world, all focused on a common endeavor. Take the European Southern Observatory, a consortium of sixteen European nations all using a series of powerful telescopes located in Chile, and a partner in a further collaboration linking Europe and Chile with Canada, Japan, South Korea, Taiwan and America. Or consider the Sloan Digital Sky Survey. Its telescope is in New Mexico, but its affiliated institutions come from fourteen countries spanning East Asia, South America, Australia and Europe. The Indian Institute of Astrophysics, headquartered in the state of Karnataka with telescopes in Tamil Nadu and Ladakh, overseas in Brazil and on the African island of Mauritius, receives a constant flow of astronomical visitors from across the globe.

But this is not a twenty-first-century phenomenon. The quality of data that emerges from an observatory has always relied on the skills and expertise of the astronomers working at the instruments. Cross-fertilization of approach—a diversity of worldviews—leads to better science. And Jai Singh, a talented mathematician and observer himself, knew this better than anyone.

Astronomy in the eighteenth century took place in three main cultures: Islamic, Hindu and Christian, which mirrored Indian society at that time. Jai Singh was a Hindu king, but his state was controlled by the Islamic Mughal Empire, run then by the emperor Muhammad Shah and descended from the Timurid Empire of Central Asia. At the same time, Christians, particularly from Europe, were widely represented in India. The historian Virendra Nath Sharma remarked:

> When Jai Singh contemplated his ambitious program in astronomy, Europeans were everywhere in the country. European travelers, soldiers of fortune, entrepreneurs, quack-doctors, missionaries, and traders were a common sight. European factories, trading posts, and settlements were scattered all over the country.[2]

Jai Singh's Jaipur observatory was originally staffed by experts drawn from the Hindu and Islamic schools of astronomy. But this left a gap.

Staffing his facility with the world's finest astronomical talent was crucial to Jai Singh's success, because he knew that astronomers elsewhere were working on bold new research programs pushing the boundaries of astronomical knowledge. He had Hindus and Muslims. What he needed to complete the set were Christian astronomers, so he threw the full weight of his prestige as an Indian ruler behind a mission to attract some.

The idea of recruiting European scholars to his research team had come to Jai Singh during the 1720s, after a few years of observing at his Delhi facility. There, he later reported, he had heard that "observatories had been constructed in a foreign country, and that the learned of that country were employed in the perfection of this important work."[3] What he had encountered was work carried out at the Paris Observatory, founded in the 1660s, particularly the astronomical tables drawn up by the astronomer Philippe de La Hire later in the century, published in Latin in 1702.

Recognizing an opportunity to flex his diplomatic muscles as well as to bring a fresh outlook to his own research, in 1727 Jai Singh commissioned a Portuguese priest working in nearby Agra, Manuel de Figueredo, to head a research trip to Europe.

Two years later, in January 1729, de Figueredo arrived in Lisbon with two companions, Pedro da Silva, an Indian-born Christian astronomer living in Goa, and (probably) Sheik Asadulā Najūmī, an Indian-born Islamic astronomer already working in Jai Singh's observatory. After an initial audience with the Portuguese king, João V, and his court mathematicians, the rest of the party's two-year trip to Europe has gone unrecorded, but by the summer of 1731, with construction of the new city of Jaipur and its monumental observatory well under way, they were back with Jai Singh to report on their findings.

The first thing Jai Singh realized when he was handed a copy of La Hire's astronomical tables was that he would need to have them translated from Latin into Sanskrit and Persian for his existing astronomers. For this, he commissioned Joseph du Bois, a French physician living in India. But there was a problem. As soon as he saw the first of du Bois's translations of the tables in 1731, Jai Singh discovered that the data they contained, and those of other tables he had acquired, did not match his own observations. By then, he had also come across reports of the work of the astronomer John Flamsteed, the first to work at a royal observatory founded in Greenwich, near London. Flamsteed had died in 1719 but his comprehensive star atlas was only published in 1729, owing to a series of disputes over the terms of publication.

Frustrated by the tables he had received from his European delegation, and eager to find out more about the latest research emerging from European observatories, Jai Singh told du Bois that he "longs for someone to go to Paris and London to drink of astronomy at the source."⁴ He decided to launch a second mission, this time approaching two French priests, Jean Pons and Claude Boudier, working in Chandernagore, 800 miles to the southeast of Jaipur, for advice on why La Hire's figures did not match his own.

Evidently, time passed slowly in eighteenth-century astronomical circles. Already stung by the four years it took de Figueredo and his colleagues to complete their mission to Europe, Jai Singh was forced to wait a further three years, until 1734, before Pons and Boudier arrived in Jaipur from Chandernagore to meet him, and when the priests did arrive, the two sides did not hit it off. By then, Jai Singh had purchased his own copy of Flamsteed's star atlas, giving him access to the latest British astronomical knowledge just as his great Samrat Yantra sundial was being constructed, and at a time when the staff of astronomers he employed at Jaipur had swelled to more than twenty.

Yet Jai Singh had not given up his ambition of attracting European scholars to live and work at Jaipur itself. In 1737, the king tried a third approach, this time inviting two mathematically trained Bavarian missionaries working in Goa, Anton Gabelsperger and Andreas Strobl, to visit him at the palace. It was an equally ill-fated trip, taking three years, before finally, in 1740, the two mathematicians arrived in Jaipur. Jai Singh must have felt he had finally succeeded in attracting the European talent he had longed for since the 1720s, but the situation quickly unraveled. Exactly a year after arriving in Jaipur, Gabelsperger died; two years after that, so did Jai Singh, aged fifty-four. With him died support for an international astronomy program across five national observatories.

But perhaps his monumental instruments had already served their purpose. With his observatory-building program, Jai Singh was taking his place in the club of great rulers. And it was an international, defiantly multicultural club.

This early-eighteenth-century network sounds vividly modern. By gathering astronomers from the Islamic, Hindu and Christian traditions, Jai Singh hoped not only to increase the quality of his knowledge but to demonstrate his place at the heart of a global network. By establishing a multicultural scholars' village at Jaipur, Jai Singh wanted his scientific experts and advisers close at hand and he hoped they would learn from each other's astronomical traditions. But like the observatories themselves, his scholarly campus was also a visible and powerful show of strength.

Jai Singh was an intensely ambitious ruler who had spent the first twenty years of his reign engaged in military conflict with neighboring states in order to consolidate his power and expand his territory. Building the new capital city of Jaipur from scratch was itself an unprecedented act. By constructing the five great observatories, Jai Singh was showing the true extent of his ambition. Time and astronomy held the secrets of the heavens. Knowledge of those secrets conferred the highest possible status since, it was thought, it demonstrated divine cosmic authority to rule, and could predict the future. Jai Singh claimed that he was descended from the Sun itself. His instruments, including the world's largest sundial, were a way to crown himself as a Sun king. He was not only authorizing and cementing his place in history, and ensuring his own political survival, but positioning himself and his rule at the heart of the universe. And he was not the first ambitious ruler to have done so.

———

IN THE WEST, the Mongol ruler Hulagu Khan may not be as well known as the empire's founder, his grandfather Genghis, or his brother Kublai, immortalized in Samuel Taylor Coleridge's famed poem written in 1797. But in the Middle East the memory of Hulagu's atrocities still looms large. In 1258, he led the sacking and destruction of Baghdad, a massacre that many see as ending what is known as the Islamic golden age, and his forces went on to invade and seize Syria. In 2002, in the run-up to the Iraq War, the al-Qaeda leader, Osama Bin Laden, broadcast a statement claiming that the US leadership had "killed and destroyed in Baghdad more than Hulegu of the Mongols."⁵

But developments in his family's internecine succession war after the siege of Baghdad saw Hulagu Khan's power under threat, and he found himself needing to show the world—and his peers—that he was a ruler to be reckoned with. Clocks provided an answer. The year after the siege, in 1259, he founded the Maragha observatory, in the East Azerbaijan province of today's Iran, and it became a science powerhouse that set a trend for high-status astronomy lasting centuries.

Maragha observatory was built on a scale rarely seen before. Its buildings covered an area of nearly 500 by 1,100 feet, with a central circular observing structure measuring 70 feet across, with numerous smaller buildings nearby. It was well stocked with instruments and a library of 400,000 books. Its quadrant, a device built into a wall for measuring star positions and tracking the passage of time, was said to have a radius of 130 feet. Constructed solidly from stone, the observatory was a powerful statement of permanence, literally rooted in the ground of Maragha and visible to the local population for miles around. Most importantly, it was staffed by a team of the best astronomers that could be found from across the vast regions of Persia, Syria, Anatolia and China.

The astronomical tables published from timekeeping and star observations made at Maragha were used around the world for centuries. Astronomical knowledge signaled control over nature and the future; command over the heavens, Hulagu Khan believed, translated to power on earth over his rivals. In 1276, his brother Kublai Khan set up an observatory near Dengfeng in China to measure time and star positions and cement his own status as founder of China's Yuan dynasty as he struggled to complete his defeat of the preceding Song dynasty. Astronomy in thirteenth-century Asia was a family matter.

Hulagu Khan had started a long-term trend. Over 1,000 miles to the east of Maragha lies the city of Samarkand, in today's Uzbekistan. In the early 1420s, the Timurid ruler Ulugh Beg, grandson of the empire's founder, Timur, established an observatory at Samarkand that became the most celebrated and influential scientific institution of the Islamic world. It was based on the Maragha observatory, which Ulugh Beg had visited as a child, as his father, Shāh Rokh, moved the family around vast areas of today's Iran and Central Asia, consolidating his hold on the Timurid dynasty following Timur's death in 1405.

Like Hulagu Khan before him, young Ulugh Beg grew up in a world of conflict and a family war for dynastic succession; he knew what it meant to gain and lose power. He also understood the power of knowledge. His observatory at Samarkand followed the lead of Maragha by using massive stone structures to reduce movement and vibration, and to allow for finely divided scales on the astronomical instruments, which were used to measure time and star positions. Like Maragha, Samarkand's astronomical quadrant was colossal; it was supported on a three-story structure that rose 130 feet into the air, faced with white marble slabs intricately engraved with marks separated by just one millimeter.

The data produced by the astronomers at Samarkand, drawn from across Asia, were translated into numerous languages and spread throughout the world. Latin translations were published in Britain in the seventeenth century, making them available for use by astronomers in the new observatories then being founded across Europe. John Flamsteed, at Greenwich, expressed the opinion that Ulugh Beg ranked equally highly alongside his European near contemporaries, including Nicolaus Copernicus and Regiomontanus. Flamsteed used his own personal copy of Ulugh Beg's astronomical tables extensively.

We can discern a pattern forming. Two powerful, expanding empires—Mongol and Timurid. Two ambitious, though potentially vulnerable, leaders. And two great observatories established. Why did Hulagu Khan

and Ulugh Beg want the latest astronomical observations at their finger-
tips? An important answer is that the predictive qualities of astronomical observation—astrology, very broadly speaking—were at the heart
of military strategies to conquer lands, and to fight wars of succession.
Hulagu Khan made Maragha a thirteenth-century center for scientific
knowledge because it gave him power. Ulugh Beg did likewise at Samarkand in the fifteenth century for the same reason.

And it happened time and again through history. In eleventh-century
Kaifeng, in China, a hugely expensive astronomical clock, set by careful astronomical observations, gave the newly installed emperor a
powerful tool to wield as he struggled for supremacy in China's ruling
parties. Its symbolism was clear: the emperor was the center around
which the empire revolved. In the sixteenth-century Ottoman Empire,
a newly enthroned ruler set up an astronomical observatory and science
program at Istanbul, in today's Turkey. Who else but a powerful and
well-connected sultan could put up the money—and gather the best
talent—to make such an ambitious scheme succeed? Brutal wars with
neighboring states followed, as this was an emperor with his eyes firmly
set on expansion.

Every empire was at it. In 1665, less than ninety years after the Istanbul facility was founded, astronomers lobbying for an observatory in
Paris wrote to the twenty-six-year-old French king, Louis XIV, urging that "the most powerful monarch in Europe" should not allow his
country to "give in to foreigners," who were then performing the most
astounding astronomical work. "The glory of your Majesty and the reputation of France," they said, rested on the construction of a great observatory.[6] Two years later, the Paris observatory was founded, a palatial
stone building at the heart of the French capital.

Like all those astronomical emperors before him, Louis founded his
observatory at a time when he was embroiled in bitter struggle; the same
year the observatory was established, Louis, who believed his right to
rule had been decided by God, embarked on a punishing half-century
campaign of war against the other major powers of Europe. Also like his

predecessors, Louis wanted the prestige that came from attracting the finest astronomical expertise to his observatory from outside his own country. Christiaan Huygens, the Dutch astronomer who invented the pendulum clock, Ole Rømer, from Denmark, and the Italian astronomer Giovanni Domenico Cassini were all enticed by Louis to work at the observatory. In 1669, Cassini was appointed its director, and became Louis's personal astronomer and astrologer for the next forty-three years.

Rivalry among rulers led to the famous observatory at Greenwich, too. After the British king, Charles II, heard, in 1674, that a French astronomer had apparently developed a foolproof way to navigate safely at sea using astronomical observations, he rushed to establish his own equivalent of the Paris observatory. It was built the following year on the site of a former castle in the grounds of Greenwich Palace, in which the monarchs Henry VIII, Mary I and Elizabeth I had all been born and lived. The observatory sat at the top of a steep hill, overlooking the River Thames, teeming with the ships of a maritime empire, visible for miles around as a powerful statement of scientific knowledge and of imperial might.

Knowledge is power. Francis Bacon said this in 1597, but it was a well-known aphorism long before that. In the Old Testament book of Proverbs, readers were told that "A wise man is strong; yea, a man of knowledge increaseth strength." A similar sentiment was expressed by Imam Ali in the *Nahj al-Balagha*. This, then, gives some context to what Jai Singh was doing in Delhi and Jaipur from the 1720s until his death in 1743. Setting up a state observatory, with an international staff and the best instruments, was what ambitious imperial rulers across time, space and cultures had been doing for centuries—and still do.

———

IN 2019, THE UK's Jodrell Bank radio observatory was classified as a UNESCO World Heritage Site, according it the same status as the observatories at Dengfeng, Greenwich, Jaipur, Maragha and Samarkand. Ten years earlier, when I was taking my turn as a steward for the Worshipful

Company of Clockmakers, I had joined forty horology specialists for a behind-the-scenes tour of the Cheshire observatory. What we saw was not a sundial or a clock. It was a time *machine*.

Sitting in the quiet countryside midway between Manchester and Stoke-on-Trent, Jodrell Bank's observatory had been founded in 1945, in the aftermath of the Second World War, to search for new knowledge using wartime radar technology. The iconic radio telescope known as the Lovell Telescope, which now defines the public image of the observatory, was set working in the summer of 1957. Its vast circular dish, measuring 250 feet in diameter and weighing a colossal 3,200 tons, steerable on circular railway tracks and with the ability to tilt at the most gymnastic angles, looms large over its sleepy surroundings.

Standing with my colleagues in the shadow of this astronomical leviathan during our horological field trip was an experience none of us have forgotten. We were all aware that observatories provide us with valuable knowledge about our place in the universe. But there was something more—something visceral—about seeing such an enormous and incongruous structure close-up. We could *feel* its power.

Like its illustrious and celebrated predecessors, Jodrell Bank Observatory is a site for gathering the most profound scientific understanding of the universe. It does not measure the passage of time, as the earlier institutions did; that work is now carried out elsewhere. Jodrell Bank acts, instead, as an observatory to look back in time by picking up the incredibly faint radio signals emitted by astronomical bodies far away in the depths of the universe. Those signals began their long journey billions of years ago, which means we are "seeing" them as they were in the distant past.

But Jodrell Bank had an earthly purpose, too. Like its predecessors, it was built and operated in the context of war. As the horrors of the global war that saw the founding of the observatory gave way to the uncertainty and paranoia of the Cold War between the Soviet Union and the West, Jodrell Bank found itself involved in an episode that was to reshape the second half of the twentieth century. Just days after it

Lovell Telescope at Jodrell Bank Observatory, photographed in June 1957

was commissioned, astronomers operating the Lovell Telescope used its radar to track the path of the rocket that put the Soviet Union's Sputnik satellite into space, in October 1957. It was the world's first artificial satellite, and it launched the Space Age.

Sputnik was a powerful weapon in the Cold War, showing the world that the Soviet Union's science and technology were supreme—and could reach into space. In the UK, it was Jodrell Bank that became a focus for the public's interest in this world-changing event. Bernard Lovell, director of the facility, claimed afterward that "it seemed to me that the eyes and ears of the whole world must be on Jodrell Bank."[7] There were grave fears, around the world, that Sputnik might be a nuclear missile.

After the drama of the Sputnik episode faded away, Jodrell Bank settled down into a program of astronomical observations that continues to this day, and about which I learned so much during my clockmakers' visit there in 2009. But for a decade and a half after Sputnik it had continued to lead a double life, offering a missile and spacecraft tracking service to both sides of the Cold War. And it was in high demand.

The US Air Force's ballistic missile division took over part of the site for a ground tracking station for the Able-1 lunar probe, planned to launch in 1958. The US Army used Jodrell Bank to track its Pioneer lunar missions in 1958 and 1959, and to send control commands to the Pioneer 5 lunar mission in 1960. During construction of a US ballistic-missile early-warning system in the early 1960s at RAF Fylingdales, deep within the North Yorkshire Moors National Park, Jodrell Bank's radio telescope was used as an interim tracking facility, ready to detect a Soviet nuclear attack. In 1962, the observatory's telescopes were being used by the US Air Force to track Soviet radar signals emanating from a host of strategic military sites, including the nuclear test range on the Kamchatka Peninsula, to the northeast of Korea and Japan, which twenty-one years later witnessed the destruction of Korean Air Lines flight 007 and the loss of all its passengers and crew.

The Soviet Union itself also made use of Jodrell Bank's radar tracking capabilities, from its 1959 lunar probes Luna 2 and Luna 3 through to its Venus and Mars probes in the 1960s and its robotic Moon landers into the 1970s.

It would be easy to see Cheshire's famous postwar observatory as a site for pure scientific research, but, of course, this is a fiction. There was great political capital to be gained in gathering knowledge of the Soviet space program (and being seen to do so), and Jodrell Bank's activities in the autumn of 1957 played a critical role in the way the space race of the mid-twentieth century played out. Placing a human in space was, of course, an act of imperial exploration; placing human boots on the surface of the Moon was an aggressive act of colonization, whatever the Outer Space Treaty of 1967 might have said. If the space race was part of an epic struggle between two superpowers fighting proxy battles above the Earth's surface to prove which leader was the most mighty, then this Cheshire observatory in 1957 played as distinctive a role in imperial power politics as did the observatories in Kaifeng in 1086, Maragha in 1259, Samarkand in 1425, Istanbul in 1577, Paris in 1667, Greenwich in 1675 and Jaipur in 1735.

ASTRONOMY IS NOT a pure science; no scientific activity is. Let us return to Greenwich, to the observatory founded by King Charles II in 1675 and which today is one of the most famous observatories in the world. Why is it so famous? At a diplomatic conference in Washington, DC, in 1884, the meridian line passing through Greenwich's principal time-finding telescope was designated the prime meridian of the world. It became the origin of longitude measurements, and of the world's time system. Greenwich Mean Time became the prime time for the world; the Greenwich observatory became the self-styled "home of time." And why was it selected for this status? Because of the power of empires.

Ultimately, Greenwich became the origin of the world's time and space because the British Empire had colonized large parts of the world and did so using its maritime power. The nation that rules the waves can rule the world, and that relied on safe navigation. As we will see in the next chapter, that had meant solving the longitude problem, which is why Charles II set up the Greenwich observatory in the first place. In 1672, just three years before the foundation stone of his new observatory was laid, Charles had founded the Royal African Company of England and bankrolled it with a substantial investment. His brother, James, was its governor and largest shareholder. The role of the company was to operate a maritime gold and slave trade between West Africa, the Caribbean and North America, and it held the monopoly rights to set up trading posts from the city of Salé in Morocco, on the northwest coast of Africa, all the way down to the Cape of Good Hope, on the continent's southern tip.

But this was about more than money and trade. Charles founded the Royal African Company just twelve years after his restoration to the throne following the British civil wars that saw the fall of the monarchy and the execution of his father in 1649, when young Charles was eighteen years old. Once he had regained his position, it is no wonder he demanded that his astronomers solve the longitude problem, because his ambitions, in Africa and far beyond, were global. He wanted glory. He wanted Britain to take over the world.

6

Empires

Observatory Time Ball, Cape Town, 1833

It happened every night, as regular as clockwork. The year was 1833, and Thomas Henderson was Britain's government astronomer at the Cape of Good Hope observatory, overlooking Table Bay on the southern tip of Africa. He had only been in the job for nine months and already he hated life here. The weather on this exposed stretch of the African coastline was unpredictable and harsh. Henderson's health was poor, and he felt let down by his astronomical assistant and his government supervisors. There were deadly snakes to contend with, and he disliked having to employ black servants at the observatory, considering them "savages."[1] Like his predecessor, Henderson preferred slavery.

But he had his nightly duty to perform, and it was time to get out his gun. In the darkness of the late evening, Henderson climbed up onto the roof of the observatory building and took out of his pocket the finely made chronometer that he had set precisely to time from his astronomical observations. In his other hand he grasped the large brass-barreled pistol and settled down to watch the hands of his chronometer tick ever closer to the advertised hour. Down toward the coast, his assistant, William

Meadows, was waiting patiently beside a time ball, a large sphere held at the top of a wooden mast that was visible to sailors on the ships anchored in the nearby bay. Meadows had his foot looped through a rope connected to the latches that held the ball up and, through a small pocket telescope trained on the observatory site, he watched Henderson intently.

Finally, the moment arrived. As soon as the second hand on Henderson's watch reached the hour, he fired the gun into the air, causing an explosion of gunpowder and a bright pistol flash that Meadows could see clearly through his telescope, immediately releasing the time ball with his foot. Down in Table Bay, the navigating officers watching keenly from the decks of their ships, having received this government time signal, were able to set their navigational instruments correctly to time and prepare to sail. This happened every night, without fail: an act of imperial timekeeping shot over the heads of the African people who were being displaced from their land and robbed of their freedom and humanity. But though he hated the apparent isolation of the Cape, and returned to Britain a few months later, Thomas Henderson was proud to play his role in Britain's growing prosperity. With his clocks and his gun, he was helping the ships of the British Empire move defiantly and freely across the globe.

———

THE ECONOMIST ADAM Smith said that the discovery of a sea route between Europe and Asia via the Cape of Good Hope was one of the two "greatest and most important events recorded in the history of mankind" (the other was the so-called discovery of America).[2] He was writing about what made nations rich, and his book *The Wealth of Nations* was published on March 9, 1776. Fifteen days later, John Harrison died, having invented the marine chronometer.

Southern Africa was one of so many places in the world where the maritime expansion of Western empires was keenly, repeatedly and violently felt. The Khoikhoi, San and Bantu people at the southwestern tip of the continent were visited by Portuguese colonizers in 1488, who named it the Cape of Storms, but it was later renamed the Cape of Good Hope by the Portuguese king, who was optimistic that it would open the riches of Asia to European trade. The first outsiders to settle at the Cape were Dutch sailors who arrived in 1652 looking for a supply base and ended up founding a colony. The British invaded in 1795, in a move against the French, with whom they were then at war; gave it back in 1803; then re-invaded in 1806. By then vast numbers of indigenous African people had been killed, dispossessed or forced away. Today, Cape Town's Table Bay is most famous for the Robben Island prison that held Nelson Mandela from 1964 for eighteen years of his twenty-seven-year imprisonment. When the British government seized control of the colony in 1806, Table Bay was one of the most strategically significant places on Earth, because the ships of every empire stopped there for supplies during their long voyages around Africa, trading the riches of imperial expansion.

But Table Bay is no safe harbor. Offering little shelter, facing due west toward the fury of the South Atlantic Ocean, its waters are whipped by ferocious storms and treacherous currents, and hidden rocks captured many careless or unlucky crews. An 1804 account described what could happen:

> arrived at the mouth of Table Bay, and just coming to anchor, a violent south-east wind rushed suddenly down from the hills over Cape Town, and nearly swamped the vessel. She was instantly laid down on her side, the quarter-deck guns driven furiously overboard, and the sails in a moment split to pieces.[3]

Yet there was no alternative, because suitable ports were in short supply on Africa's southern coastline.

Navigating safely round the continent demanded the highest skill and the best technology and, by the time the British planted their flag on the Cape, the best navigational technology meant marine chronometers, highly accurate timekeepers first developed by John Harrison in the

1750s, that offered a fixed time reference from which longitude could be calculated once at sea. Crucial to the process of chronometer navigation was access to accurate time signals during stays in ports, so that officers could check their shipboard chronometers and set them to time. And in an age before radio, these coastal time signals themselves needed access to accurate time, either from a nearby astronomical observatory or, as electric telegraph networks began to be built from the 1840s onward, from a telegraph time service that linked back to an observatory else- where. Time signals and chronometers kept empires afloat.

The first regular time signal at Cape Town was a cannon fired daily at noon from a battery on the castle, which had been built in the seven- teenth century on the Table Bay waterfront. The castle had allowed Dutch settlers to protect their new colony, but in 1806, when the Cape was seized by British forces, the incoming military engineers installed their latest and most powerful guns there. The noonday firing, which began the same year, was an expression of military power as much as it was a practical navigational signal, although it would be easy to argue that the two pur- poses amounted to the same thing. However, the time which the gun signaled each day was not, to begin with, accurate enough.

The Cape was formally handed over by the Dutch to the British in 1814 but it took until 1820, as Britain's empire grew, for the idea of a perma- nent astronomical observatory and time signal station overlooking Table Bay to come to its Board of Longitude, so that navigators could get a bet- ter time fix for their chronometers before setting sail. The site chosen for the new Royal Observatory, on high land to the northeast of Table Moun- tain with a commanding view of the Bay, missed the worst of the clouds and sand dust that covered sites elsewhere in the colony, but it was a challenging place for the newly appointed astronomer, Fearon Fallows, who had grown up in the relative comfort of Cumbria and Cambridge. "In the days of Fallows," wrote a later Cape astronomer, "this site was part of a bare, rocky hill, covered with thistles, infested with snakes . . . the jackals howled dismally around it at night, and a guard of soldiers had to be established to protect the property from theft."[4] It was also a burial ground for enslaved people and the wind could sometimes be so strong as to knock riders from their horses.

It took a few years for the permanent observatory buildings to be completed, but even while he was working out of a temporary hut in the garden of a rented cottage in town, Fallows started providing time signals for the ships in Table Bay. From 1821 he sent his assistant, James Fayrer, with the correct time to the docks so that captains could check their chronometers against Fayrer's corrected watch. Two years later, Fallows began shining a bright oil lamp toward the water each evening before shuttering it at a fixed time. Fallows died in 1831, replaced by Thomas Henderson, who set up the time ball service down in the bay two years later, triggered by his pistol flash from the new observatory roof. When he quit the observatory soon after starting the time-pistol service, he promised to tell friends back in Britain "all about my residence in Dismal Swamp among Slaves and Savages."[5]

From the moment Fearon Fallows had arrived at the Cape, he had built the newly founded observatory on slavery. In 1822, he boasted to his British government supervisors of an "experiment" in which he would apply for and train a series of enslaved boys in observation techniques so that he would eventually build up a workforce of astronomical assistants that "would always be under *absolute command*."[6] In 1825, as building work on the observatory began, the contractor advertised for "sixty good Masons, and as many Labourers, to work at the Royal Observatory; likewise, thirty strong Slave Boys by the month."[7] Fallows also employed formerly enslaved people at the observatory as servants, but preferred direct slave ownership where possible. In 1831, Thomas Henderson commented that "Of course a certain number of servants is indispensable—but after the experience which has been already had, great trouble and vexation will be saved by purchasing Slaves at once."[8] Fallows had originally forced his enslaved workers to sleep underground in cellars, before reluctantly converting an instrument storeroom into sleeping accommodations.

Slavery was abolished in 1834 but the astronomers' exploitation of indigenous Khoikhoi people as servants continued. An observatory

Cape observatory with time ball on mast, engraved in 1857

assistant, Charles Piazzi Smyth, who arrived at the Cape in 1835, claimed
in later life that "the badness, dearness and scarcity of native servants
out there, is one of the chief troubles of Cape life."[9] The book in which
Smyth recorded punishments given to local workers was literally termed
his "Black list."[10]

In 1836, Henderson's successor, Thomas Maclear, set up a time ball in
the grounds of the observatory itself to replace the pistol signal. Later,
in the 1850s, a new time ball at the Cape Town docks gave sailors tied up
in Table Bay a better view, and time-disc signals were set up at Simon's
Town, on the peninsula overlooking False Bay, and at Port Elizabeth and
East London in the Eastern Cape. From 1861, all the balls and discs were
operated by an electrical signal from the observatory, as was the old
time gun on the Cape Town castle. The electrical signal came from a
battery in a wooden hut at the observatory. East London was 600 miles
from Cape Town. In 1893, a time ball was set up at Port Alfred, halfway
between Port Elizabeth and East London, and an electrically controlled
clock was set up in the Harbour Tower at the Cape Town docks giving
observatory time to the second. The following year, a new time ball was

erected on the waterfront nearby, which is still there. All this infrastructure was built just in order to distribute accurate time from the observatory to the chronometers on board each of the ships in the bay. Time on the southern African coast must have been worth a fortune.

———·———

THE EARTH TURNS on its axis once each day, and that is why clocks and trading empires are inseparable. Trade means travel. Overseas travel. Over seas. The problem comes when you can no longer see land. Where are you? Where is the place you need to sail toward? What perils lie between you and your destination? Global seafaring empires based in Europe and built on overseas trade began to grow in the fifteenth century. Portugal started capturing people and land in 1415. Spain followed in 1492, then France from about 1534 and the Dutch Empire began to form in about 1543. Britain was late to the party, starting to seize land in 1583 but only really getting going twenty years later. The British politician and colonizer Walter Raleigh, in about 1615, said that "whosoever commands the sea commands the trade; whosoever commands the trade of the world commands the riches of the world, and consequently the world itself."[11]

Sailors became remarkably adept at navigating on long ocean voyages. Measuring the height of the Sun or stars could give them their latitude, or north–south position. A compass could give them their direction using the magnetism of the Earth. Log-lines—ropes with evenly spaced knots paid out over the side of the ship—could give a good estimate of speed when used with an hourglass. The saltiness of the water; the presence of different types of flora and fauna; the direction of currents and winds; the depth of water near coastlines; and meticulous record keeping on good sea charts all helped solve the problem. But there was a big one that remained, which was a reliable method to find longitude, or the ship's east–west position. Without it, crews risked crashing into rocks or, just as fatal, running out of supplies when land was further away than they thought. Often, it just meant inefficient voyages in which sailors traveled more slowly than necessary or took long diversions to ensure safe arrival.

The "flying Dutchman," a legendary tale from the eighteenth century, told of a ghost ship that haunted the Cape of Good Hope in bad weather, unable to make the harbor and doomed to roam the seas for eternity. Its origins came from seventeenth-century Dutch East India Company ships sailing around Africa on their way to India and Southeast Asia.

A typical voyage involved sailing from Europe straight down the middle of the Atlantic Ocean until the navigator's instruments indicated the ship had reached 34 degrees south, the latitude of the Cape, before turning east and hoping that a few days' further sailing would bring land into sight. That is how the first Dutch settlers reached the Cape in 1652. But often enough ships would miss their target and overshoot, having to sail north until meeting the southeastern coast of Africa before turning back, slowly hugging the coastline until Table Bay came into view. On particularly ill-fated voyages, the ship might bounce around, back and forth, trying to reach a haven in furious weather, losing time, food and fresh water as it did so. As disease and thirst took more and more sailors, eventually the crew could lose control of the ship and it would become lost, carrying only the corpses of those last to die. Another flying Dutchman would haunt the oceans around Africa.

But as European maritime empires grew ever larger and more ambitious, there was no time to beat about as there was money to be made. Loss of life at sea was regrettable, but loss of valuable cargoes hit profits, and money was therefore the real motivation (of course) behind solving the problem. The answer to the problem of longitude, as everybody had known by the middle of the sixteenth century, was clocks, because the Earth turns on its axis once each day.

———◦———

THE LOCAL TIME by the Sun in Cape Town is different from the local time in Mumbai by three hours, thirty-seven minutes. The time difference is exactly equivalent to their longitude difference, where each hour of time is equal to fifteen degrees of longitude. So, Cape Town and Mumbai are 54.5 degrees of longitude apart. All you need to find your longitude position when at sea, relative to a fixed place such as your port of departure, is the local time at that port, and at the place where you now are, at the same instant.

Finding the local time on board a ship in the middle of the ocean was quite straightforward. When the Sun reached its highest point, it was local noon. Devices that measured angles solved this: cross staffs, back staffs, octants, sextants. Finding local time at the port you had just sailed from was the hard part. It seems obvious now that sailors simply needed to carry a clock on board set to port time (let us say Cape Town). In fact, this was known as early as 1530, but at that time there did not exist a clock or watch on Earth that could keep time to the accuracy, precision and stability needed during a months-long ocean voyage, with the violent movement of the ship, the extremes of temperature, and damage caused by the salt- and water-laden air. This was a problem for clockmakers.

But there was a clock *above* the Earth that could do the job instead, if only its time could be read. As early as 1514, experts started drawing up plans to use the Moon and stars in the sky like the hands of a giant celestial clock. This was called the lunar-distance method, because it used the angular distance of the Moon from certain stars on any given night to tell navigators the time at a fixed place on Earth. The trouble was, the stars had not been mapped systematically enough for this to work, and, anyway, nobody had yet built a portable device to measure the angles accurately enough. This was a problem facing astronomers and instrument makers.

A mechanical timekeeper or the lunar-distance method. Either would give sailors the time at their home port, but neither was yet possible, so the big maritime nations started throwing money at the problem. In 1567, Philip II of Spain put up a cash reward for a practical longitude scheme and his successor, Philip III, fronted another one in 1598. No success. In 1600, the Dutch government founded a longitude reward scheme. There

were lots of proposals, but nothing worked. In 1667, the French government set up the Paris Observatory as part of its finance minister's plans
to increase trade and build France's overseas empire, hoping an astronomical solution to longitude would soon follow. It did not. In 1675, the
British government followed suit, establishing the Royal Observatory,
Greenwich, to map the stars and solve the longitude problem. Lots of
valuable work was carried out but no solution was yet forthcoming.

Time passed, and more and more ships were lost. Two thousand people were killed in 1707 when a fleet was wrecked on the Scilly Isles, off
Britain's southwest coast. It was one of Britain's worst maritime disas-
ters; more people died that night than were killed on the *Titanic*. The
pressure to solve the problem of maritime navigation once and for all
was growing ever stronger and from all quarters—from merchants, from
ministers, from captains and from the public, whose families were being
decimated by the relentless loss of life at sea.

In March 1714, during her speech at the state opening of Britain's Parliament, Queen Anne exclaimed that "this Country can Flourish only
by Trade; and will be most Formidable, by the right Application of our
Naval Force."[12] The following month a group of merchants and sea captains responded to her call with a petition to the government, in which
they said that:

> nothing can be either at home or abroad, more for the common
> Benefit of Trade and Navigation, than the Discovery of the Longi-
> tude at Sea which has been so long desired in vain, and for want of
> which so many Ships and Men have been lost.[13]

The answer, they said, was a reward scheme, and just three months later
the government passed an act founding a Board of Longitude that offered
£20,000 to anybody who could solve the problem to half a degree. It
took half a century for the British prize fund to bear fruit, and then it did
so twice. By the late 1750s, John Harrison had completed a mechanical
timekeeper (later known as a chronometer) that kept time well enough

to win the reward money and coped with the rough conditions at sea. By the same time, a consortium of astronomers and instrument makers had mapped the stars and developed the sextant and nautical almanac that made the lunar-distance method a practical reality.

It took a few decades for these two methods of finding longitude to become commonplace. Lunar distances took a lot of effort every night, while people were naturally wary of trusting their lives to a mechanical timekeeper. Some captains took a lot of convincing that a chronometer was worth the expense, but officers involved with pioneering chronometer voyages from the 1770s onward increasingly extolled their virtues. The first captain to use chronometers extensively was the East India Company's James Dundas, who commanded an East Indiaman to India in 1779. As the ship went around the Cape of Good Hope, Dundas found that the longitude he had calculated using the chronometer was almost spot on, and he used the devices enthusiastically on future voyages.

Slowly but surely, the supply of chronometers went up, the price came down, and resistance to the new arrival was chipped away as the demand for profit in an expanding world market inexorably rose. By the early years of the nineteenth century, navigation had been transformed, with both chronometers and the lunar-distance method in widespread use and, as Queen Victoria came to the throne in 1837, the race was on for imperial supremacy as maritime nations battled to build the biggest global empires.

———

THE ROYAL OBSERVATORY in Greenwich was responsible for testing all the British Royal Navy's chronometers as they came off the production line or passed between ships, and it had 500 machines on test at any one time, all in a room about the size of a small studio flat: "what a wonderful instance of the proof of our maritime power is this apartment!" exclaimed the astronomer Edwin Dunkin in 1862. "On entering, the visitor is startled by a universal buzz, which sounds almost like the hum of a beehive."[14]

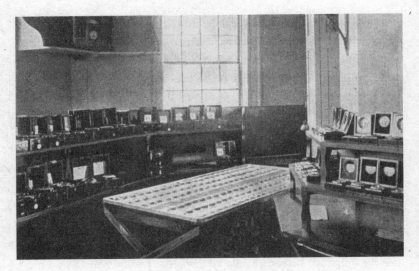

Royal Observatory, Greenwich, chronometer-testing room,
photographed c. 1897

It still retains that magical presence today. When I worked at the
observatory as its curator of timekeeping in the 2000s, the former chronometer testing room became my office, a room into which all the Royal
Navy chronometers serving the British Empire had been carried for
their crucial tests for over a century. What a remarkable century. I could
almost hear the echoes of their soft, regular, faithful ticks. I almost for-
got what they had enabled.

The observatory was just one of many sites where chronometers flowed
in and out like the tides. Chronometer makers themselves received their
products back many times during their long lives for servicing and over-
haul. For them, accurate time shown on the workshop clock was vital
to get their machines beating the perfect seconds on which sailors' lives
and the prosperity of empires depended.

Before electricity brought time signals to street corners from the 1870s
onward, the only way chronometer makers could know the time accurately enough for their needs was either to measure it themselves, using

a portable time-finding telescope called a transit, which was difficult to do in reality, or to send their shop assistant to their nearest observatory with a pocket watch to ask for the time. Even the best chronometer makers tended to prefer weekly observatory visits over using a transit. But the constant interruptions at the observatories from people with watches asking for the time started to become a nuisance from the 1820s onward as chronometers came into widespread use.

At Greenwich, the problem was addressed by switching things round. From 1836, instead of waiting for the chronometer trade to visit, an observatory assistant, John Belville, started taking a corrected pocket watch round the London makers, charging an annual subscription for a weekly visit with a timekeeper accurate to one tenth of a second. When he died in 1856 his wife, Maria, took over the business, and when she retired in 1892 their daughter, Ruth Belville, continued the weekly visits until 1940, when London was in the depths of the Second World War.

It is likely that similar networks existed in other places around the world where chronometers were made—Liverpool, Paris, Hamburg, Copenhagen—though their stories are yet to be pieced together. But it was one thing to distribute accurate time around big cities. It was quite another to get it to all of the empire's countless ships tied up in places like Table Bay, far from the imperial capitals, so that the chronometers on board could be set right before the ships set sail. The time signals standing around the coastlines of the world were foot soldiers at the frontiers of empire.

———

IN 1908, A survey carried out by the British navy listed a total of 200 time signals—usually balls, but sometimes discs, guns or flags—that could be found on coastlines or in ports around the world, and the document reveals the astonishing reach of Europe's maritime powers. But it also speaks volumes about the shifting sands of global geopolitics.

There were twenty time signals around Britain's coast itself, from

Edinburgh in the north to Falmouth in the south, and a further fifty-one had been erected by that date in British colonies overseas, including installations in India, Singapore, Africa, Australasia, Canada, Malta, Gibraltar, Mauritius and the West Indies. It was by far the world's biggest network of time-signaling stations, but by 1908 America was fast catching up as a world power, having built twenty-two coastal time signals on its own soil, with a further two in its port possessions in the Philippines.

The states of Norway, Denmark and Sweden, between them, had twenty-three time-signaling stations, but only around their own treacherously convoluted coastlines. Germany, France and Russia each had ten signals on the 1908 list, although only four of these were mounted in overseas territories: one in Germany's imperial naval base at Tsingtao, China, and three in France's possessions in Africa. By this time, the overseas empire of the Netherlands was focused on Indonesia, where two signals had been built, and Suriname, on South America's northeastern coast, where there was one time disc. A further signal was available on the island of Curaçao, off Venezuela's coast. There were four signals erected around the coastline of the Netherlands itself.

What of the other big European empires? By 1908, Portugal's overseas possessions had dwindled to the point where the country operated only four time signals—in its colonies in Africa and the Azores—as well as one in Lisbon. Spain, having once been the greatest imperial superpower, had recently lost much of its last remaining territory in the Spanish-American War of 1898–99, and in 1908 only operated time signals at its home ports of Vigo and Cádiz. Austria-Hungary had built four signals on its lands in today's Croatia and Italy, while Italy itself had come much later to the overseas imperial scramble, although it had, of course, long been a proud maritime nation. In 1908, it had no overseas signals but had built a network of eight around its own long coastline—including a time ball and gun at Catania, the Sicilian city that had provided ancient Rome with its first public timekeeper over 2,000 years earlier. A further twenty-one time-signal stations dotted the coastlines of China, Japan, Egypt, Cuba and Hawaii, with five more around South America.

Yet, while this geopolitical atlas of time signals can help shine a light on the shifting sands of maritime empires at the turn of the twentieth century, it can never reveal the whole story. Belgium only merited one entry in the 1908 time-signal register, a set of four circular discs mounted on the tower of a government building in the port of Antwerp, despite the fact that it had been engaged in the most vicious imperialism in the heart of Africa for over a quarter of a century. But until that year it had been a private colony, not a state one. In the 1870s and 1880s, Belgium's king, Leopold II, had worked with the British colonialist Henry Morton Stanley to privately occupy what is today the Democratic Republic of Congo. For twenty-five years he made a personal fortune from rubber, ivory and minerals extracted from his possessions. But the sickening cruelty his agents perpetrated on the people of the region became hard for the international community to ignore. As evidence of the scale of atrocities mounted, in 1908 the Belgian state reluctantly took over the king's private holding, giving Belgium a major overseas colony for the first time. Yet all this is missed if we simply look at the single time signal in Antwerp.

Nonetheless, whatever the individual histories of empires, this global infrastructure of navigation does give us a vivid sense of the overall scale of the imperial project as the twentieth century came into view. Each and every one of these time-signal installations was a major practical and legal undertaking, and involved complex engineering, scientific research, technological development, land acquisition, the recruitment of labor forces, legal negotiation, agreements, protocols, maintenance, record keeping, astronomical observation, instrument making and a quite extraordinary amount of expertise developed over decades or centuries. Each one was a heavy investment and a long-term liability and, taken together with the tens of thousands of shipboard chronometers for which they were built, and the global network of chronometer-testing stations, makers, retailers and supply depots, not to mention the astronomical observatories that found the time in the first place, we can see an astonishing global infrastructure of time—of clocks in one form or another—that have been all but forgotten. Plot all these time signals and chronometers on a map of the world and you can see how clocks enabled empires.

Or, better yet, look at the clocks that showed off empires for all to see and marvel at—and still do. In 2020, I was invited on a special visit to see behind a door in London's Piccadilly Circus Tube station that is normally kept firmly locked shut. Keith Scobie-Youngs, who runs the Cumbria Clock Company and looks after some of the UK's most important public clocks, had asked if I wanted to see an early electric-clock mechanism, hidden in a cramped cupboard on the station concourse, that he had been asked to refurbish. I was quick to accept. As I joined Scobie-Youngs in the little enclosure, I found a Heath Robinson–like assemblage of wheels, pinions, levers, motors, wires and electrical coils, which looked for all the world like a child's Meccano dream. But as I came to understand what did what, and which part was connected to which mechanism, I realized it was a different sort of dream. It is the sophisticated device that drives the station's world-time clock, a map of the world with a brightly illuminated horizontal band at the equator marked with the hour numerals of time that slowly rotates throughout the day and night. These days, passengers hurry past the clock without noticing it, intent on reaching the platforms deep below and onward to their destination. But when the clock was first set running, in April 1929, it showed the dream of a global empire, and it was besieged by onlookers.

World-time clocks like this show the time anywhere on Earth. All you need to do is find your place on the map and follow a vertical path with your finger to the illuminated equatorial band. The numerals there will tell you the time. But this one offered a little more than that, as London Underground's 1920s managers picked out six places for special attention. Each one of the six was marked with a tiny light bulb and was connected to the time band by a polished metal spike, sharpened at the end like an arrow. But it could equally have looked like a stake, driven hard into the Earth to signify occupation and ownership. A news report in 1928 described the newly built Piccadilly Circus Tube station as "the Hub of the Empire," and passengers ascending its new escalators were

Piccadilly Circus world-time clock being demonstrated in 1929

greeted by a huge painted mural of the world—with Britain firmly at its center.[15] The world-time map installed on the station's circular concourse should have looked like a chart of Britain's imperial success. But when we scrutinize the six places picked out for the public's attention we can see an imperialist's dream turning into a nightmare.

Greenwich sits dead center, of course: the seventeenth-century observatory that had been founded to serve an expanding empire by a king running an African slave-trading operation with his brother. Greenwich, the observatory that by the twentieth century had become the very center of time and space. But what of the other five spotlit cities?

The next place to be picked out with a light bulb and a metal stake on the clock is New York, which the slave-trading king and his brother had seized a few years before they built Greenwich its observatory, but which had broken away from the clutches of the British in such a decisive fashion a century later. Near it are the lamp and stake marking the city of Victoria, in Canada's British Columbia. It was named after the queen who had colonized it in the nineteenth century, but by the 1920s it was almost free of British control. It was a similar story below the map's illuminated equator. The fourth lamp on the map is Sydney, in Australia, a land that, like Canada, was finally sloughing off the British. La Plata, in Argentina, is the fifth lamp. This was no former British colony but had been the focus of intense trade diplomacy between Argentina and Britain in the 1920s—its lamp was symbolic of the new empires being forged that favored new worlds, not old ones. And the sixth lamp? It was Cape Town, and it, too, was finally throwing off the yoke of the British imperialists.

By the mid-1920s the imperial map of world power was changing fast. The US economy and naval strength were on the rise. Japan, Italy and Germany were all shifting into a high gear. By contrast, Britain lost control of both Egypt and the Irish Republic in 1922, and control over all its dominions was weakened in the 1926 Balfour Declaration. By the 1930s, South Africa moved away completely and India began moving toward independence. But the crowds at Piccadilly Circus Tube station in 1929 saw none of this. In this bright electrical clock, mounted proudly on a marble wall surrounded by lavish bronze fixtures, the traveling public saw Britain, or rather London, at the center of the world, and they saw the sharpened metal stakes of imperial ownership planted firmly and defiantly in Africa, the Americas and Australia. It was all a fiction by then, but it was a fiction many people still believed. Because what happened in Piccadilly Circus was happening on a vast scale in another hub of the 1920s British Empire. With clocks, the message of British supremacy was being transmitted not just to a Tube station but around the world. And this message was emanating from a field in the Warwickshire countryside.

———

BY THE 1910S a new technology was offering something that had once seemed like a miracle: long-distance communication without wires. Who needed to wait in port for the daily time signal when wireless came along? Sailors could check their chronometers every day while at sea, which meant they could go faster because their navigation was more accurate.

The first international time signals by wireless were broadcast from the Eiffel Tower in Paris from 1912. Ten years later a network of fifty transmitting stations around the world provided daily time signals for ships at sea. All the major maritime empires were served: there were wireless stations in British India, the Dutch East Indies, French Indo-China and Portuguese East Africa, among many others, but none in Britain itself, much to the consternation of Frank Hope-Jones, chair of the Radio Society and a leading manufacturer of electric clocks, who said in 1923 that "the Englishman, who regarded Greenwich time as something peculiarly British, has been getting it from the observatories and countries of his neighbours to an increasing extent for the last ten years."[16] But plans were afoot to right this wrong.

Rugby radio station was opened by the General Post Office in 1926 as part of Britain's Imperial Wireless Chain, a project set up after the First World War to connect the whole of the British Empire together by wireless communication. It specialized in long-wave transmission, which could travel over very long distances as its radio waves followed the curvature of the Earth rather than radiating in straight lines and ending up in space, and was designed to communicate with the entire Royal Navy at once, reaching as far as Canada, India, South Africa and Australia directly from the UK. One of its first jobs was to transmit twice-daily time signals for navigation, under the call sign GBR.

It would be easy to see the GPO's Rugby radio station as a technical solution to a communication and navigation problem. But it meant far more than that. As the British Empire was rapidly falling apart, Rugby was built as part of a government public-relations campaign to boost Britain's sense of imperial pride, and it was impossible to miss the GPO in the 1930s. Its film unit made distinctive and innovative programming for cinemas around the country. It developed promotional services like a telephone talking clock. It invited newsreel cameras to the opening of

every switchboard, exchange and radio station. But what it was really
doing was trying to prop up the very idea of a united British Empire—or,
rather, of Britishness itself.

In 1932, the GPO welcomed the Pathé news company to Rugby to film
"The World's Greatest Radio Station." In one scene, as a Morse buzzer
chatters away and its message spools out on a printed paper tape, viewers
learn that:

> whilst we are here, this transmitter is sending out news to liners in
> all the seven seas. No place is inaccessible to GBR. For those who
> can't read Morse, this machine is signalling that the Post Office
> Rugby Station is the most powerful in the world.

What this scene was *really* signaling was that the British Empire still
considered itself the most powerful in the world. The radio station's
transmitters were merely a proxy for the empire, and the film was made
as a rallying call to the British people. But it was too late. After the Second World War ended, Britain was on its knees as America and the USSR
became the world's superpowers and each one geared up for the Cold
War. One by one, Britain's former possessions cut their ties with the
mother country, with India's independence in 1947 a particular blow to
Britain's imperialists. But empires cast long shadows, and we still live
with them today.

The clock that was originally installed at the Greenwich observatory
in 1927 to transmit the time to Rugby for broadcast to the British Empire
was a type called a "free-pendulum," which actually comprised two
devices, one called the "master" and the other termed the "slave." The
master clock was free and did little work; the slave clock was forced to
march to the master's beat and carried out all the labor of time distribu-
tion to the radio transmitters. The terminology of slave clocks had been
coined in 1904, by a British government astronomer, but not at Greenwich.[17] It was in Africa, at the Cape of Good Hope observatory, at the
height of the Western imperial scramble to carve up the continent and

its people, that a white British official, working at an institution whose very walls had been built by enslaved people, chose to enslave clocks themselves. Over a century later, people still routinely use the racist terminology of master-and-slave systems in engineering and horology, yet it carries a violent weight of the imperial past, born in Africa.

In 1949, the original GBR time signal from Rugby was joined by another, known as MSF. The MSF signal is still broadcast today, though the radio transmitters moved from Rugby to Cumbria in 2007. When I was curator of timekeeping at the National Maritime Museum, I helped preserve the last ever Rugby MSF clock just before the service moved out. The clock had been built as recently as the 1990s and comprised a rack of electronic equipment the size of four wardrobes side by side, and contained three independent sub-clocks, each with its own chain of driving circuitry to the transmitter: one cesium atomic clock, one rubidium atomic clock and a third clock set right by GPS satellites. Three independent drive chains increased reliability and allowed the three clocks to be compared with each other, improving accuracy.

One carried a large plaque labeling it the "red drive chain." One was captioned the "white drive chain." And one carried the words "blue drive chain." Red, white and blue. The white chain carried an additional sign that designated it the "master."

The writer John Agard was born in 1949 in the Caribbean country known then as British Guiana. It was part of the British Empire until 1966, when it gained independence and took the name Guyana. Agard moved to the UK in the 1970s. In 2007, he was working at the National Maritime Museum as writer-in-residence and I showed him the Rugby MSF clock which we had just acquired. A few days later, he wrote a poem about it, entitled "The Rugby Clock":

Today, if I may, I'll tackle the Rugby Clock.
Sorry if rugby fans are suitably shocked
that I choose to scrum with the language of clocks.
I promise not to mention the All Blacks

But rather how metaphors transmit their tracks
onto Empire's timebound timepieces.
How Britannia dwells in ticktock spaces
where radio-ruled time governed the dusky races.

Allow me to kick off with the red, white and blue
which is familiar as rhubarb stew.
Observe the red drive chain
the white drive chain
the blue drive chain.
All masters in their own relentless right.
All masters and equal in the sight
of cosmic time's timeless curriculum.

But since some masters are more equal than some,
—if I may borrow an Orwellian dictum
without putting too fine a point on it—
note which drive chain takes the master credit
and how the master frequency control overseers
the—wait for it—white drive chain.

Now, I hear you ask, where does blackness fit
into this time-dictated continuum?
Am I, so to speak, winding you up for a scrum?
Blackness as mother of time will be my next thesis.
Meanwhile, time ticks its nemesis—
For aren't we all chained by the wrists?

Manufacture

Gog and Magog, London, 1865

For over a month, the building at numbers 64 and 65 Cheapside had been missing its entire façade and roof while a major refurbishment and remodeling took place, and five stories of complex scaffolding shielded the site from curious bystanders on this crowded City of London thoroughfare that leads from St. Paul's Cathedral to the Bank of England. John Bennett was expanding his clock- and watch-making business, and he knew that to survive meant to innovate, so he was turning his Cheapside premises into a tourist attraction, filled with the latest horological technology that would draw onlookers—and customers—from far and wide. It was a Friday morning in May 1865, and the project was in its final fortnight. For the building contractors, this was a lucrative job. As a retail remodeling, it was unusual not just in its scale but in the nature of the work being carried out, and, that day, many of the workers were at the very top of the building helping the foreman install a huge piece of specialist equipment behind the roof parapet. It was heavy and difficult work, made harder by the cramped conditions on the highest level of the scaffold and the unseasonably warm weather

London was enjoying. Everyone was tired, hot and fatigued from a long week when suddenly, without warning, disaster struck.

All the workers on the Cheapside site, and the passersby on the busy street below, heard the foreman cry out in fright as the scaffold board on which he was standing, at the top of the five-story building, gave way beneath him. It happened so fast that nobody could move, and all they could do was watch in horror, helpless, as he plunged headlong toward the ground. Death seemed certain.

As he fell, his arms flailing, the foreman passed in front of the colorful, sculpted figures of two giants, named Gog and Magog, the traditional guardians of the City of London, who were standing ready to strike the quarter-hour bells of the huge clock above the entrance to John Bennett's remodeled horological emporium. Gog and Magog had been installed by the foreman and his workers just days before, and they would go on to become one of the most popular tourist attractions of London, drawing crowds so great that a permanent police presence would be needed to keep the busy street moving. But all that was in the future. On this hot Friday morning in 1865, a more awful scene was unfolding in front of the eyes of Gog and Magog, and those of all the shocked onlookers in Cheapside.

Then a miracle happened. In the foreman's moment of need, something saved him from a terrible end. Perhaps it was the protective presence of the guardians Gog and Magog, or maybe it was simply good fortune. In his panic, as he hurtled downward toward the pavement below, the foreman grabbed wildly for one of the scaffold poles that formed a lower stage of the structure, caught it, and somehow managed to use his downward momentum to swing himself inward, where he landed heavily, but safe, on the thick wooden planks. It was an "extraordinary escape from death," as one newspaper described it the following day, and, in the statement he gave after the incident, while still in shock, the foreman told colleagues he could not believe he had survived, believing a "horrible death" was certain.[1] But survive he did.

This incident happened at a time when John Bennett was at the height of his career as one of Britain's most famous and influential watchmakers. It also took place when the world of manufacture—how things are made, where in the world, by what sort of people and to whose benefit— was being profoundly transformed. Those looking for allegory might see, in the foreman's near-death experience at a watchmaker's shop in London, a parallel with this wider shift. By the 1860s, one of Britain's oldest and most venerated craft industries was, itself, hurtling toward what seemed like certain death, and Bennett was trying desperately to save it.

JOHN BENNETT HAD been born into a two-generation watchmaking family in Greenwich in 1814 and joined the family business as an assistant when he was about fifteen years old, after his father died. Working alongside his mother and a small staff of watchmakers, he spent the next few years learning his trade by repairing horological mechanisms from all manner of customers, including repair work contracted out by the nearby Royal Observatory, a ten-minute walk from the Bennetts' premises. He got to spend time with the country's top astronomers, and there must have been worse lots in life than to earn a livelihood from the study of horology while living in Greenwich. But it was not enough. By his early thirties, the ambitious and energetic Bennett was itching to strike out on his own account and, in 1846, he set up his own business in Cheapside, at the heart of the City of London on a vibrant shopping thoroughfare. The firm he established stayed on that site for decades after Bennett himself retired in 1889.

Victorian Cheapside was a sensational place to visit. "New York may boast her Broadway," exclaimed one 1865 newspaper article, "and Paris her Boulevards, but Cheapside beats them hollow in the great essential of the sensation element—a steady stream of population on foot, and in every variety of vehicle, from early morn till dewy eve."[2] Bennett's firm itself later claimed that Cheapside was "the very centre of our great City's trade, in fact it has been more—it has been the very heart of the world's commerce."[3]

Bennett was no ordinary watchmaker. In an industry populated by modest, often introspective characters, the person in Cheapside stood out with his vigorous embrace of modern publicity, marketing and advertising techniques. In 1851, with his horological business still a fledgling, Bennett paid the unheard-of sum of £750 (over £100,000 today) to get a full-page advertisement on the back of the catalogue of the Great Exhibition, held that year in a gigantic crystal palace in London's Hyde Park. Eleven years later, he paid £1,000 (nearly £125,000 today) to buy a front-page space in the catalogue of the International Exhibition of 1862. He made it all back and more.

Bennett's 1865 remodeling of his Cheapside premises demonstrated his grasp of public relations for all the world to see. And the press went wild:

> Gracious goodness! What a sensation was in Cheapside the day the enclosing screens were taken away! 'Busmen came to a standstill; the wild hansom was paralysed; railway vans moved not; telegram boys climbed up lamp-posts; the footways were literally packed with people to look at the new "hammermen."[4]

The "hammermen" were Gog and Magog, and when the fateful scaffolding came down in June 1865, visitors were faced with a dazzling cornucopia of horological gadgets bolted to the fascia and displayed prominently in the wide new front windows.

At the very top of the building was the device that had nearly been the death of the construction foreman: a towering five-foot-diameter red time ball on a twenty-foot mast, just like those being erected by the score around the world's imperial coastlines to follow the lead of the original version at the Greenwich observatory, set up by the Astronomer Royal, John Pond, in 1833, and the rudimentary time ball inaugurated the same year at the Cape observatory. Bennett knew Pond personally from his Greenwich days, and would have known the Greenwich time ball inside out—not just technically, but in the powerful statement it made to the world from its commanding position overlooking the River Thames and

Gog and Magog behind a projecting clock on John
Bennett's shop front, with other horological attractions,
photographed in 1891

London's crowded docks: here is progress. Here is modernity. Here is the *future*. By 1865, time balls had become potent symbols of Victorian values in the high-tech world of electric telegraphs and split-second timing, and John Bennett's ball, which dropped every hour triggered by a signal from the Greenwich observatory itself, was the third to appear in central London and could be seen up and down the famous Cheapside.

Underneath the time ball, set into a large recess at the building's third story, was a large, sonorous hour-bell for the massive turret clock driving the shop's horological displays, struck by a life-sized automaton winged figure of Father Time carrying his hourglass. In the story below that was a second niche, this one holding the giant, brightly painted automaton figures of Gog and Magog, the guardians of the City, each emerging every fifteen minutes to strike a quarter-hour bell. In front of them was a huge projecting clock dial, brightly illuminated with multicolored gas jets at night, proudly proclaiming the name of "Bennett" along the busy thoroughfare as far as the eye could see.

Gilded signs plastered over the fascia claimed Bennett was "Watch Maker to the Queen" and "Clock Maker to the Royal Observatory," though as far back as 1849 the then–Astronomer Royal, George Airy, had told his assistant to sever all ties with Bennett, claiming in exasperation that "I will have no more to do with him."[5] Bennett proudly described his premises as the "City Observatory," which must have grated with the punctilious government astronomer. His relationship with Queen Victoria and her staff, however, was much warmer, by all accounts.

At street level, beneath the gaudy façade of hammering giants and colorful gas jets, was a display in the shop window that showed how well Bennett understood the rhetorical power of precision in the Victorian age, instilled in him by his work for the Greenwich observatory in the early years of his career. At Cheapside, like any clock or watch shop, he had a large clock known as a regulator, a timekeeper like the grandfather clocks in countless domestic hallways, but of the finest quality and offering accuracy and precision good enough for astronomical use. Normally, a regulator would keep time to something like a few seconds in a week. But Bennett wanted better. The same electrical signal that triggered the time ball on the roof every hour also arrived, or "flashed," at his shop's regulator, where it performed an exquisitely fine duty. A newspaper report, probably ghostwritten by Bennett himself, described the mechanism as follows:

> The flash will act upon a portion of the mechanism of this regulator in the shape of a triangular wedge, which will drop into a notch in a circle, in the centre of which revolves a hand to mark seconds of time. So nicely is the mechanism arranged, that, should the regu-

lator by any chance be in the slightest degree wayward in its movements, even to the hundredth part of a second, the electric current will correct it.

This public service, readers were told in glowing terms, was "what may virtually and even literally be called the 'exact' time."[6] The significance of this in Victorian culture cannot be overstated.

John Bennett's shop became a public sensation. It was so popular with onlookers and tourists that Cheapside was habitually blocked by people waiting for the next automaton display. It was once described as being "only second to the Strasburg crowing cock as the most famous and

Crowds on Cheapside, outside John Bennett's shop, photographed in 1904

popular horological variety entertainment in the world."[7] The obstruc-
tion was so bad that soon after the building was revealed, one of Bennett's neighbors, the Manchester Insurance Company, took him to court,
accusing him of ruining their business with his new-fangled marketing
contraptions. The two sides bickered back and forth for a few days in the
local newspapers, with the magistrate describing Bennett's displays as
"a piece of tomfoolery" and Bennett retorting, rather grandly, that they
were "horological science."

But, despite the bluster and puffery, Bennett had an important point
to make:

> It is not the business of men in authority to be the first to snub
> enterprise and improvement in any branch of trade or commerce.
> They would be far more likely to show their worth when they strive
> to give full play to every man's ingenuity in his own particular pro-
> fession. By this the world would be a constant gainer.[8]

Bennett's self-publicity and modern methods stuck in the throats of the
conservative elite. He was too pushy; too willing to cast aside the conventions of the past; too full of himself. But the world was changing.
What role was there for innovation in Victorian manufacture? With his
gadgets and slogans, and his delight in provoking the anger of the conservative establishment, this maverick showman was powering into the
future, and his exuberant new shop display shouted this fact from its
very rooftop. But this was not the only reason why the British watchmaking industry hated him so much.

Bennett was not famous just for his modern approach to marketing and his taste for publicity. By the 1850s he was also renowned—or
notorious, in some eyes—for his outspoken views on Britain's horology
industry and for his tub-thumping public lectures delivered to packed
audiences up and down the country. There was one further display forming part of his Cheapside shop front, which had appeared in his window
as early as the 1840s, that told of the true significance of his pioneering

but controversial approach to manufacture. It was a small sign that simply advertised the sale, by Bennett, of timepieces imported from France, Switzerland and America. But this little sign made the clockmakers mad as hell. In the conservative, insular and closed world of British horology in the middle of the nineteenth century, to sell foreign-made clocks and watches was tantamount to heresy. Yet here was John Bennett, the most public watchmaker, proudly showing off his heretical practices for all to see. And what happened in horology stood, in one way or another, for every other manufacturing industry.

THROUGHOUT THE EIGHTEENTH century, Britain had been the world's biggest maker of clocks and watches. By the end of the century, its watchmakers were producing over 150,000 watches per year, many for export overseas, which was about half of all world production at the time. Its horological industry, based primarily in London, Coventry and Liverpool, had evolved into a finely organized network of specialist crafts- and businesspeople working together to produce fine-quality, though quite expensive, timepieces. They were expensive because the skills were hard-won, and the practices and hand tools passed from one generation to the next in an old guild-style apprentice system. But it had been buoyant.

So, when the pioneers of the new textile industry of the 1760s, such as Richard Arkwright, went looking for skilled workers to make complicated cotton-spinning mechanisms for the huge factories, or mills, that they built, they approached England's clock- and watchmakers. In 1768, Richard Arkwright went into partnership with the clockmaker John Kay to build the first example of the spinning machine that went on to change the world of manufacture, patented the following year. Kay worked with a watch-tool maker on parts of the pioneering machine. Arkwright's textile innovations took off rapidly and, two years after the spinning machine was patented, Arkwright advertised for specialist machine makers for his new spinning factory in Derbyshire: "Wanted immediately, two Journey-man Clock-Makers, or others that understands Tooth and Pinion well."[9]

Very soon, the new textile industry exploded into life, with textile mills springing up primarily across Lancashire. As it grew, commented the textile industrialist John Kennedy in 1815, "a higher class of mechanics, such as watch and clock-makers, white-smiths, and mathematical instrument-makers, began to be wanted."[10] In an industry far removed from their own, the clockmakers of England were changing the face of manufacturing—and the modern world.

It was not just the textile mills. All the industrialists had been at it. When Josiah Wedgwood, the pottery entrepreneur, wanted machine tools for his new factory in 1767, he approached the clockmaking trade. When the mechanical engineer and steam-engine maker James Watt needed tools for his Birmingham works, it was to the clockmakers he went. When his business partner, the manufacturer Matthew Boulton, was asked for advice on who made the best machines, he recommended the clockmakers of Lancashire. Two decades later, the civil engineer John Rennie complained to Boulton that the cotton trade had deprived London of its best clockmakers, leaving the London trade deficient.

The very term "clockwork" became synonymous with industrial machinery. The word was used to describe the mechanisms of spinning machines, with some pioneering devices said to be of a "delicacy equal almost to that of clocks," and others "moved by clockwork," forming "an aggregate of clock-maker's work and machinery most wonderful to behold."[11] When the cotton mill designer and textile machine maker Peter Atherton died in 1799, one of his warehouses was found to be full of clock- and watchmaking tools.

This was the story of the eighteenth century. There had been little incentive to find new ways to manufacture clocks and watches because there was a home and export market happy to buy everything the British manufacturers turned out. And even if trade were to slacken off a little, there was ample work for talented clockmakers in the textile mills of Lancashire or the machine shops of Birmingham. But as the eighteenth

century gave way to the first decades of the nineteenth, things started
to change, fast. And John Bennett had been one of the first to catch on.

———

IN DECEMBER 1859, John Bennett stood in front of a meeting of the newly
formed British Horological Institute and proceeded to tear into Britain's
education system. In Switzerland, he exclaimed, every child of every
class, girls and boys, in the most remote mountain districts as well as
the major cities, was given the best education. They were taught mathematics, languages, science and mechanics. But it went further. "Beyond
this," Bennett commented, "a special regard was paid to the cultivation
of all that would refine the taste or elevate the character of the future
man or woman."[12] All children were taught music, design, drawing and
art. "Who could wonder, then, that Swiss watches were found, not only
cheaper in cost, but immensely distancing our own products in beauty
and elegance?"

Then he turned his focus on to the assembled crowd of watchmakers
working in Clerkenwell, London's horological heartland.

> Surely it is high time for Clerkenwell to take a leaf out of the Swiss
> book; and this might be done if the members of the Horological
> Institute would earnestly support those members of the trade who
> refused to shut their eyes to the existence of facts by which alone
> our prosperity could be secured.

And, with that, he delivered his manifesto for the salvation of British
watchmaking. "Nothing would now avail but an entire change of system,
the employment of the female hand, and the highest possible cultivation
of the rising generation."

It was not the first time he had told the clock- and watchmakers of
Clerkenwell they were on the way out. Two years earlier, after a stormy
public meeting, one maker had stood up and accused Bennett of "an

array of falsehood, mockery, and perversion."[13] It is fair to say he was preaching heresy. But everything was clear in Bennett's mind. The Swiss system of watch manufacture was the only way forward; the only way the industry could be saved. And he spent his career urging a five-point plan.

First, watch mechanisms must be simplified. Second, measurement should be standardized and decimalized. Third, machines should be used to make repeatable parts to the finest tolerances, rather than workers using hand tools to shape and fettle each part to fit. Fourth, division of labor should be taken yet further and women brought into the workforce in large numbers. And, fifth, the elementary-education system, as well as the technical schools, must be transformed in quality and scope. All this would bring costs down without sacrificing quality, just as the Swiss had achieved.

But the Brits would not listen to the heretic from Cheapside. As early as 1842, the London trade had had the chance to import Swiss methods wholesale when the Swiss entrepreneur Pierre Frédéric Ingold attempted to set up a watchmaking factory, fitted out with a range of specialist machines, in Soho. In a parliamentary debate, one MP exclaimed that "Nothing was impossible to machinery in the present day. When they saw what had been accomplished by steam, how could it be asserted that anything was impossible to be affected by the ingenuity of man."[14] But the scheme came under sustained attack from the indigenous trade, and it failed. It did not take long for the British watchmaking industry to fail also. By 1870, Switzerland was producing over two-thirds of the world's watches by value and the British trade was on its knees.

It is easy to point the finger of blame with hindsight and from positions of comfort. But it is hardly a surprise that there was resistance to radical change. Yet problems do not disappear if they are ignored. For the whole of the nineteenth century, the trade refused to believe that foreign manufacturers could do a better job than they did, and they carried this prejudice with them as the industry twitched and kicked in its death throes. As late as 1889, one Clerkenwell old-timer defiantly wrote, "I for one fail to see the good of this high-flown technical learning. Keep the foreigner out, that is my motto."[15] It was always somebody else's fault.

But Switzerland was not the only foreigner parking its artillery on Clerkenwell's lawns, for there was another challenger waiting in the wings, ready to take on all comers. A new system of manufacture had been developing slowly through the nineteenth century and, by the 1870s, cheap watches and clocks, mass-produced in factories using specialist machines powered by steam engines and operated by low-skilled workers, were rolling off the production lines that formed part of the system.

The old ways of manufacture were disappearing, trumpeted a British watchmaking journal in 1875, to be "superseded by the disciplined organization of the factory, the youthful hand, the automaton machine, and the steam-engine! Even Switzerland itself is threatened by the all-conquering power."[16] The new conquering power was America; the new system became known as the American System of Manufactures; and while clocks were its first products, the system changed the face of manufacturing, worldwide, forever.

———

IN ABOUT 1802, a maker of wooden clocks in Connecticut made a business decision that changed his life and the course of history. Eli Terry decided to manufacture hundreds of clocks at a time, and market them himself, rather than constructing small batches of a few clocks to order. He built a small factory to house water-driven machines to make them. And they all sold. Five years later, he started manufacturing a batch of 4,000 clocks in a new factory that took him a year to set up. These sold, too, and proved to doubters that it was both possible and profitable to mass-produce clocks. Then Terry took matters a radical step further. In 1816, he patented a new, more compact wooden clock that was not only mass-produced from machine-made parts, but from parts that were interchangeable with each other. Nobody had done this before; it was the start of a manufacturing revolution.

After four years, Terry's factory was turning out 2,500 of these clocks each year, with just thirty workers. By 1830, Connecticut boasted twenty-three clock factories operating on the Terry principle, producing 38,000 clocks per year between them. Terry's clocks were simple affairs. But his workshops acted as an incubator for innovation, and the people who learned production techniques from him went on to develop the idea in exciting ways.

Terry sold his first factory to two of his pupils, Silas Hoadley and Seth Thomas, in 1810. Thomas developed it into a system of machine tools. Then, in 1837, another of Terry's pupils, Chauncey Jerome, started manufacturing a batch of 40,000 clock mechanisms—made from brass, making them smaller and more reliable than Terry's and Thomas's wooden clocks. They were also cheap: a dollar and forty cents per piece. Jerome built a second factory, and between them his two facilities were churning out 280,000 cheap brass clock mechanisms annually. By 1840, he was looking for overseas markets to soak up the supply and decided to try exporting a few to Britain. Soon they were selling there by the shipload at prices much lower than their British-made equivalents, and quickly became popular in rural communities. Somerset farmers, for instance, were happy to trade in their lumpen old English grandfather clocks for a colorful, compact new American clock, and farmhands working for low wages could imagine, for the first time, proudly owning their own clock. Little clock shops began springing up in rural areas where, before the influx of cheap American clocks, there had simply been no viable market.

With such massive production figures, advertising played a huge role in realizing the success of this American-led project, a fact that John Bennett in Cheapside knew only too well. The international exhibitions of trade and industry, such as the 1851 Great Exhibition in London, offered the opportunity for manufacturers to reach millions of potential customers and sales on a scale never seen before. Bennett's shockingly expensive advertisements on the catalogue covers of the 1851 and 1862 exhibitions were the very least a forward-looking manufacturer would entertain. The emerging horological factories in America mounted increasingly lavish displays at these expos, featuring thousands of watches and aggressive marketing techniques to hook in customers. When Bennett exhibited at the 1878 Exposition Universelle in Paris, he did so alongside such technological innovations as the telephone, electric lighting and the phonograph. The future was built by *selling* modernity, not just inventing it. It is hard today to imagine manufacturing without marketing, so critical a part has it become of our lives.

It has been described as an American invasion, and for the rest of the century the British trade all but gave up trying to compete in the market of cheap clocks. By the time it did come around to the American way of thinking, it had missed the boat. American clocks, and then French and German ones following the American lead, eventually killed the British clockmaking industry as surely as Swiss watches had put Britain's watchmakers out of work.

But Eli Terry's manufacturing innovations in horology, and those of the other mass-production pioneers in Connecticut and elsewhere, played a powerful role in transforming manufacture much more widely. The American System, as the world came to know it, took two ideas— standardized, interchangeable parts, and the use of special-purpose machine tools to replace skilled human labor—and unlocked the door for repeatable mass production of complex products as diverse as guns, sewing machines and typewriters.

But for this new world of mass production to grow and flourish, a new breed of heavy, highly precise specialist machine tools would be needed. The development of American machine tools in the 1860s and 1870s, suitable for widespread use in manufacturing high-precision and interchangeable parts, followed a similar pattern to the British developments exactly a century earlier which brought the factory system into existence and kick-started an industrial revolution. It also followed the pattern of developments in Connecticut fifty years after that, when Eli Terry initiated mass production. Who was already steeped in precision, fine measurement and accuracy? In order to make the machines that made machines, all eyes turned once more to the clockmaking industry.

Take Brown & Sharpe, the American machine-tool manufacturer founded in 1853 that had become the biggest in the world by 1900. Its founder, Joseph R. Brown, was formerly a clock- and watchmaker and, in the opinion of one engineer who trained in Brown & Sharpe's machine shops in the 1870s:

> deserved greater credit than any other man for developing and making possible the great accuracy and the high efficiency of modern machine practice and in making it possible to manufacture interchangeable parts . . . I know of none who deserves a higher place than, or who has done so much for the modern high standards of American manufacturers of interchangeable parts as[,] Joseph R. Brown.[17]

That engineer was Henry Leland, founder of Cadillac, who introduced interchangeable parts to the automotive industry.

With the mass production of the later nineteenth century came mass consumption, and mass marketing sold the products of manufacturing to an ever wider public. Entrepreneurs, and those able to mobilize worldwide markets—those able to move with the times—survived and thrived. John Bennett sat squarely in this category. Those who did not—like the vast bulk of the British horological industry—eventually died.

———

WHEN THE MESSAGE is unpalatable, there is a tendency to shoot the messenger, and it is no exaggeration to say that the British watchmaking industry came to detest John Bennett.

After Bennett died in 1897, aged eighty-three, the knives came out sharp. In the august pages of the *Horological Journal*, the trade paper of the watchmaking and clockmaking industry, the editor said that Bennett's personality was "peculiar" and that he was "deficient in veneration and dignity, but with abounding assurance and self-esteem." Though he bought a lot of stock from the London watchmakers, "he was not much liked by them," and "indeed no one behind the scenes ever took Bennett seriously."[18] But the public loved him, habitually crying out, "What's the time?" when he took part in civic processions as a City of London alderman in the 1870s.[19]

"The world is my country!" Bennett had made an impassioned statement in 1886 in which his allegiances were laid out for all to see, commenting that "to get the best watch for the least money, that is my business as a servant of the British public . . . why should I have watches made in England when I can obtain better goods for a lower price by employing workmen in Switzerland rather than workmen in England? You must remember that although a small class in this country may suffer, the great mass of the community benefits."[20]

The historian David Landes has claimed that the British clock and watch industry failed owing to "high costs, conservative styling, obsolescent technique, entrepreneurial complacency, resistance by labour to innovation."[21] He concluded: "The watch trade gave warning of things to come."[22] With this critique of British horological failure, Landes might also have been describing the ailing fortunes of the British automobile industry a century later. And with cars the giant figures of Gog and Magog, that perennially popular horological variety act installed on the front of John Bennett's Cheapside emporium, come back into view. Because they did not die with John Bennett himself in 1897, nor when his company's Cheapside premises were demolished in 1929 to make room for an insurance company.

———————

IT WAS JUST after eight o'clock one morning in 1928 when Henry Ford knocked at the front door of London's Science Museum, asking to be let in. The attendant on duty told him the museum did not open until midmorning, but when Ford informed him who he was, it did not take long for an urgent message to reach the museum's keeper of engineering, Henry Dickinson, who promptly raced down, accompanied by other senior colleagues, to meet the famous American carmaker and industrialist.

Henry Dickinson might have been me, but for the passage of a few decades. During the fifteen years when I was a curator—latterly a keeper—in the Science Museum's engineering collections, many was the time I would take a call from my colleagues on the front desk asking me to come and meet visitors who had arrived with questions about the artifacts in my care. It was never a chore. I have always loved discussing the history of technology with people who share my interests. It is all too easy for back-office staff in big museums to become disconnected from the people who really matter—the public. Occasionally the timing could be inconvenient if I was in the middle of something or heading for one of the meetings that seem to be a constant feature of museum life. But I always came away with new knowledge and insights from these impromptu conversations.

I like to think that Henry Dickinson felt the same sense of anticipation as he walked a path I came to know so well, striding through the museum's majestic East Hall, filled with steam engines from the eighteenth and nineteenth centuries by engineers such as Thomas New-comen, Richard Trevithick and James Watt, before bounding up the stairs into the museum's entrance hall to meet his visitor. Having ushered Ford through the doors that morning in 1928, Dickinson proceeded to tour the carmaker around the museum's rich array of exhibits, doubtless asking more questions than he answered. Later, Ford was shown the museum's engineering section, which includes such icons as *Puffing Billy*, the world's oldest surviving locomotive, and Robert Stephenson's *Rocket*, which ran between Liverpool and Manchester on the world's first modern railway. Ford was overwhelmed by what he saw, and I am not surprised. I was, too, every time I had the pleasure of showing the museum's visitors the very same exhibits.

The purpose of Ford's visit in 1928 was to scout for engineering relics for a museum he was building at the Michigan town of Dearborn, and his ambitions were as expansive as his spending. Several times during his visit, Ford offered to buy the historic exhibits, including *Rocket*, only to be politely rebuffed by Science Museum staff.

But Ford was also building a historic village alongside his museum, with reconstructions of legendary American buildings such as Orville and Wilbur Wright's cycle shop, Henry J. Heinz's family home and the courthouse in which Abraham Lincoln practiced as a lawyer. It opened to the public in 1933 and remains one of America's most popular attractions. But there is one building in the village that does not seem to tell an all-American story of innovation, and it came about during Ford's visit to London in 1928, after he had tried and failed to buy the Science Museum.

He was being driven up Cheapside and, before long, the car reached the long-familiar figures of Gog and Magog sounding the quarter hours outside the Bennett company premises. Ford was not even two years old when they were first installed, the time when the foreman nearly lost his

Henry Ford's staff removing Gog and Magog from John Bennett's shop front, 1929

life setting up Bennett's time ball on the roof above. But Bennett was a manufacturing entrepreneur and reformer after Ford's own heart, and, as he saw the famous clock, Ford immediately resolved to buy the entire shop front, its clock, and Gog and Magog themselves. Though his offer was initially resisted, this time his persistence won, and the deal was concluded the following year.

But Bennett's horological gadgetry was not about to give up its commanding position in the City of London's premier thoroughfare without a fight. Some sixty-four years after Bennett's builders had labored to install the shop front and its horological attractions, Ford's contractors struggled for weeks to dismantle it, only being permitted to work on Sundays owing to the narrowness of the street and the volume of traffic passing down. Eventually, though, Gog and Magog reluctantly gave up their positions keeping a watchful eye over London and were packed away into crates, along with Bennett's shop front and clock, and shipped to Dearborn, where they remain.

Henry Ford had started life as a watchmaker, not a carmaker. At least, that was the romantic tale the company told:

> Watch repairing was coming easier to young Henry Ford. He had started at 14 and the first watch (today in his private collection at Dearborn, Michigan) had been mended with a nail, tweezers made from a corset stay, and a pair of knitting needles . . . Years later, the watchmaker's skill and precision which young Henry Ford had learned in those winter nights were to be used in building more than 30 million cars and trucks. Moreover, it was Mr. Ford's experience with watchmaking that gave him the idea of using an assembly line in building automobiles.[23]

In his 1922 autobiography, Ford explained that during his teenage apprenticeship at a Michigan maritime engineering facility in the 1880s he worked nights repairing watches in Robert Magill's jewelry shop in Detroit. He remarked:

At one period of those early days I think that I must have had fully three hundred watches. I thought that I could build a serviceable watch for around thirty cents and nearly started in the business. But I did not because I figured out that watches were not universal necessities, and therefore people generally would not buy them. Just how I reached that surprising conclusion I am unable to state.[24]

Instead, Ford switched to automobiles, and what a switch. His motor company pioneered the form of manufacture termed "Fordism." Mass production with repeatable parts made by machines was taken to its most efficient conclusion with Ford's moving assembly lines, in which the product moves, and the workers stay still.

It had taken a long time to get to the stage of Ford's Model T, the car that motorized the masses in the first decades of the twentieth century. As early as the 1760s, the Scottish historian and philosopher Adam Ferguson had written, "Manufactures, accordingly, prosper most, where the mind is least consulted, and where the workshop may, without any great effort of imagination, be considered as an engine, the parts of which are men."[25] The automotive historian Andrew Nahum has commented that this could be a "perfect description" of Ford's factory.[26] Yet it took the convulsive developments in manufacturing of the eighteenth and nineteenth centuries, heavily influenced by clock- and watchmaking culture and championed by the likes of John Bennett, before the dream of the perfect assembly line could be realized. And Ford knew it.

Both Bennett and Ford had got into watch repair as teenagers, and both went on to transform their respective manufacturing industries in the face of deeply entrenched resistance. John Bennett may have died poor and largely forgotten in 1897, remembered only by the spiteful watchmakers of Clerkenwell. But he had made his mark. With Bennett's help, and with the protective guardians of Gog and Magog watching over Michigan, clocks and watches made Henry Ford one of the richest people on earth.

Morality

Electric Time System, Brno, 1903–6

It was a proud day for the Antel family as they gathered in St. James's, the medieval second church of the historic Moravian city of Brno, in today's Czech Republic. It was early in 1907. Three days after Christmas, Ludmilla Antel had given birth to Edith, her first child with her husband, Johann, and today was Edith's baptism. Everybody was wrapped up in their thickest clothes to keep out the winter chill, huddling together for warmth as the rest of their family and friends arrived. Outside the doors of the church, the electric trams whined and crackled as they buzzed briskly through the street. Brno's residents were still getting used to these modern marvels, which had been installed in 1900 to replace the noisy, slow, horse-drawn and then steam-driven trams that had been a feature of Brno's daily life for decades. The city's central square, a few steps from the church, was brightly lit by electric lights, pushing back the winter gloom, as was the public piazza in front of Brno's main railway station, in the far distance. These public-lighting installations, too, were new: part of the city's modernization program that had been going on for several years. Brno was a city on the move.

St. James's Church and clock tower from Brno's central square, photographed in the early twentieth century

Once the Antel family and friends had gathered at the church, the time arrived for Edith's baptism ceremony and, as if to mark the auspicious occasion, the huge public clock fitted high up on St. James's tower began slowly to strike the hour on its heavy bell. Close by, across the town square, the clock on the town hall started striking at precisely the same time. Across town, passengers hurrying to and from the railway station who happened to glance at the clocks mounted on each of the building's twin towers saw the time exactly in synchronization with the bells from the church and town hall.

It was only a few months since Brno's electric time system, covering the railway station, town hall and St. James's Church, had been completed, and it was state of the art. Out went the older, traditional tower clocks, each driven by heavy stones that fell as weights, and each telling its own idiosyncratic version of Brno time, demanding dangerous feats of acrobatics to keep them wound, working and on time. In their place came a network of sophisticated modern mechanisms, driven by powerful electric motors, impervious to the winter wind, rain and snow that could easily put a stop to their weaker, hand-wound predecessors. Electricity also meant that the mechanisms high up in the towers could get their time through electric wires fed from finely made, high-precision pendulum clocks housed in central office locations, kept dry and warm, and cared for daily (and safely) by local officials. And it meant that all the clocks could show the *same* time: as a network of cables buried under the streets flashed time signals at the speed of light to the tops of the towers, Brno's public clocks were able to present a synchronized, standard time for the city and its residents.

Modernity had come early to Brno. Today, fans of modernist buildings make pilgrimages there, paying homage to Brno's rich crop of 1920s and 1930s architecture by celebrated figures such as Ludwig Mies van der Rohe, whose Tugendhat Villa, nestled in lush gardens near the city center, is such an important example that it was designated a UNESCO world heritage site in 2001. But these modern buildings marked the end of an era in Brno, not the start of one. The electrification of Brno had begun in 1882, when officials building its new German City Theater invited the electricity pioneer Thomas Edison to design and fit an electric lighting system, instead of the gas illumination that was common at the time. The decision followed devastating theater fires in Nice, Prague and Vienna the previous year, each caused by faulty gas equipment, that had killed hundreds of theatergoers. At Vienna, the gas fault led to an explosion that set light to the fly system—the theatrical rigging—soon followed by the auditorium. The emergency lighting failed to operate, and the emergency exits only opened inward, trapping the audience in the inferno. The theater's manager ended up in prison. By switching to the latest in lighting technology, Brno's theater became one of the world's first public buildings lit by electricity, and the city soon got a taste for it.

The electric time system installed in Brno between 1903 and 1906 was part of a wave of time standardization sweeping progressive cities across the world, each with electricity at its heart. Like the electric trams and lighting schemes, it was changing the lives of those who lived in these modern cities. But electric clocks were a subtler addition to everyday life than trams and bright lighting. For most churchgoers, the striking of the new electric clock at St. James's was a familiar and welcome interjection into their day, though quickly ignored. But, as the bells started to ring that winter's day in early 1907, Ludmilla and Johann Antel glanced appreciatively up to the tower above them, before looking at each other, and down at little Edith, with fond smiles of pride on their faces.

Johann Antel was one of Brno's most innovative modern clockmakers, and when the city had wanted a state-of-the-art new electric time system to be installed in its most prominent public buildings, it had come to him. Antel had helped take Brno into the future.

———•———

WE HAVE ALWAYS believed that electricity is the future. Electric cars, electric trains, solar panels on the roof, wind turbines, the cell phone in your pocket: these are all taken to be symbols of a better, cleaner, faster, safer, more connected future. Electricity travels at the speed of light and the world is a vast, pulsing network of signals on the wires and through the ether. Electricity has come to define modernity: it *is* modern.

Electric clocks, like those installed in Brno and countless other towns and cities around the world, may seem like a mundane, practical technology that simply helps us get around more efficiently. The clocks on the walls of our workplaces, railway stations and other public places have merged into the background, and we rarely pay them a second glance. But that only serves to prove how effective they have been in changing the behavior—the very morals—of our societies since they first started to be developed in the second half of the nineteenth century. Far from being mundane, these clocks have been used for the most elevated moral purposes. For 150 years they have been tools to standardize the behavior of the masses in line with what those in power consider to be the *right* behavior. And that is because electric time systems allow time itself to be standardized.

———

GREENWICH MEAN TIME is the time on the Greenwich meridian, the line that runs due north and south through a building at the town's Royal Observatory, founded in 1675. It is the time local to Greenwich. What it is *not* is the time at Bristol, 110 miles due west of the Greenwich observatory. The local time at Bristol according to the Sun is ten minutes behind Greenwich. In 1675, each town kept to its local time. But today, Bristol, along with the rest of the UK, keeps Greenwich time, because in the nineteenth century people around the world decided to standardize time.

Standard time is the system where everybody in a city, region, country or continent agrees to set their clocks to the time in one place, such as Greenwich, which becomes the standard. For anywhere east or west of that place, the real time according to the Sun—local time—differs from the standard time, but for reasons which are all about morality and the principles of good or bad behavior, people decided that did not matter.

When people tell the story of how time became standardized in the UK, they invariably talk about the building of railways. The first passenger railways were constructed in the 1830s and 1840s and they quickly showed the need for a standard time for the whole network. Otherwise, how do you run an east–west railway like the Great Western Railway, running between Bristol and London? You would have to change your watch at every station. When passengers needed comprehensible timetables, and the safety of the system depended on time to separate trains sharing single tracks, a commonly agreed single timescale was not just convenient but lifesaving. So the local time at each station along the GWR gave way in 1840 to the railway's own time, which was chosen to be the time in London, and the only way to get the right time in London was to get it from Greenwich. This, in turn, was possible thanks to another new network

Assistant at the Greenwich observatory time-signal control desk, c. 1897

being built (literally) alongside the railways: the electric telegraph, which, as well as transmitting messages in the dots and dashes of Morse code, could also send time signals instantaneously. We saw earlier that coastal time signals around the world, set by electrical signals from nearby observatories, helped imperial navies keep their navigational chronometers on time. The same principle helped the railways grow and flourish, too.

A single clock in Greenwich could announce the moment of 10 a.m. hundreds of miles away using an electrical impulse sent automatically along the telegraph wires. The same electrical time signal could be received in every major station along the line, and local arrangements could carry the time deeper into the branch-line network, too. At each station the clocks could be set to Greenwich, and in every employee's pocket the watch provided Greenwich time on the move. With that single central standard clock, a whole rail network could be kept running on one time, and by the 1850s every railway in Britain had adopted the same practice.

This account is all well and good, so far. But most histories of technology at this point go one step further. They commonly claim that, by 1855, it was not just railway time but the UK's own *civil* time—the time in everyday life across the entire nation—that had standardized to Greenwich, and we abandoned our multitude of local civil times. That is how the story of standard time is usually told. The assumption is that everyday life quickly falls in line with the practice of the early adopters, in this case the railway companies. But the assumption is wrong.

It would have been possible for us to have two times in our lives, railway time and local time, and to convert between the two. It seems as if it would be a pain, but we manage to operate today with two time systems (twelve-hour and twenty-four-hour), which sometimes necessitates a bit of mental arithmetic. We use two length and weight systems at the same time (imperial and metric) and we have two temperature systems (Celsius and Fahrenheit). It is true that some people think all this should be done away with, but that is an argument for another day. The point is that we manage to live our lives with multiple measurement systems running in parallel. It could have been the same with local time versus standard time, and what the railway storytellers miss is that it *was* the same. Local time *did* hang around, and it did so until the 1880s, half a century after the railways started to run. The railways played a role in time standardization, but there is a much bigger story of how standard time in everyday life came about. And it revolves around Victorian morality and electric clocks.

IN 2007, I was working with James Nye, a fellow historian of electrical timekeeping, on a research project about the Standard Time Company, which had been founded in 1876 to promote an automated time-distribution network across London based on electricity. Partway through our project, James found the STC's 1886 subscriber list, together with a hand-drawn network diagram showing where its wires ran across London. And we knew immediately that he had hit gold. What we discovered from that document came to change the way we thought not just about electricity and the standardization of time but about the Victorian world, its morals and its relentless quest for modernity.

We pored over the subscriber list and wiring diagrams, tracing the details onto old maps of Victorian London. We worked out how old road layouts matched those of the present day and consulted countless street directories to find exactly where STC's wires ran. We spent time walking the streets looking for clues about the size of the market—how big the companies were and how modern their premises would have looked in 1886. And, after a few weeks of study and exploration, we started to realize what was going on.

In 1886, more than 300 customers in 336 different locations were hooked up to STC's electric time network, getting an hourly burst of electric current which could automatically correct every existing clock in their premises. Thousands of clocks on the walls of these buildings were set right, to the second, by STC's electric synchronizing signal that fired across a telegraph network every hour from clocks at STC's control center in the City of London. Tens of thousands of people relied on STC's standard time to coordinate the business of state, finance and commerce.

For many of the subscribers—such as banks, exchanges and clearinghouses—it was easy to imagine why they might need to know the right time. But there was another type of business that made up a quarter of the company's entire subscriber list. Over eighty London pubs, cafés and restaurants paid for a feed from STC's electric time network. At first, James and I could not figure out why. So, in search of clues, we set ourselves the challenge of ascertaining how many of these standard-time pubs still existed, which meant a lot of walking and, when successful, drinking. We did not drink as much as I hoped, because most of the pubs had closed long ago or been flattened in the two world wars since STC floated on the stock market. Many had been swept away in postwar

reconstruction. And, of course, we could only visit the pubs during opening hours, between noon and 11 p.m.

———•———

THE REGULATION OF the sale of alcoholic drinks was one of the UK's
most controversial pieces of proscriptive legislation of the Victorian age.
Anthony Dingle, a historian of alcohol temperance, wrote that "Victori-
ans were obsessed with alcohol. Among the literate and the articulate,
the proper place of drink in society was debated with an intensity and
an exhaustiveness which is now difficult for us to comprehend."[1] The
1872 Licensing Act was the first that introduced nationwide restrictions
on the hours of sale of liquor to the public, and the debate over just this
act alone, never mind all those that preceded and followed it, was long
and acrimonious. Issues of class, freedom, public health and the authority of the state were all bound up in it, which was a perfect example of
Victorian morality and the use of clocks to police human behavior. In
Victorian Britain, temporality enabled temperance and, by introducing
licensing hours, the state had called time on drinking.

But which time? In the 1870s, despite the railways, local time was still
alive and kicking across the UK. A crucial 1858 legal judgment known
as *Curtis v. March* had ruled that the official time for Britain's courts was
local time, not Greenwich time. This was the problem facing temperance
legislators. If you want to put time limits on the sale of alcohol, you need
two things: firstly, to agree whether any given pub was keeping local or
standard time; and, secondly, for everyone to have access to that time
accurately, because if you are going to prosecute publicans for selling
alcohol after hours, you will only succeed if the hours themselves are
beyond reproach.

In 1874, Parliament was debating how well the 1872 Licensing Act was
working in practice, and the discussion turned to the vexed question
of clocks. One MP in the debate suggested that all pubs should be connected to the Royal Observatory so that they would have no excuse for

not knowing the time. Just two years later, the STC was founded, and this explains why so many pubs paid for its time service.

But the second problem was harder to solve. Were the licensing hours in any town or city to be based on local time or Greenwich time? One MP argued that "there would be a great deal of advantage in securing uniformity of time in reference to the working of the measure" and that the law should be amended to say that official licensing hours "shall be reckoned according to the time kept at the Royal Observatory at Greenwich."[2] Six years later, the matter was finally solved when, in August 1880, an act was passed in the UK that defined Greenwich Mean Time as the official time standard for all British legislation, with Dublin Mean Time for Irish laws.

Local time passed into history not because the railways used it, but because 1870s anti-alcohol reformers wanted to use clocks to police their moral crusade. And they were not the only moralists who wanted to do so.

In 1902, the British socialist reformer Sidney Webb claimed that:

> Of all the nineteenth century inventions in social organisation, Factory Legislation is the most widely diffused. The opening of the twentieth century finds it prevailing over a larger area than the public library or the savings bank: it is, perhaps, more far-reaching, if not more ubiquitous, than even the public elementary school or the policeman.[3]

Webb was writing exactly a century after the first Factory Act, which had been passed in 1802 to help protect children working in Manchester's cotton mills and which was the first to put a limit on working hours. Numerous acts followed in the early decades of the nineteenth century, but there was a common thread that linked them all together. They all took time for granted, relying on factory owners and managers to operate the clocks that defined the working day. As you might imagine, this approach was not altogether flawless.

The 1844 Factory Act was the first to define an actual time standard that was to be used in regulating working hours. "The hours of the work of children and young persons . . . shall be regulated by a public clock, or by some other clock open to public view," and the clock needed to be approved by the district factory inspector.[4] This meant an end to the use of a type of clock which ran at the speed of the factory's machinery, rather than showing the true time, and therefore cheated workers when the waterwheel driving the machines was running slowly. But it was clear that the new standard of factory time remained open to abuse.

The 1880 Act that defined Greenwich Mean Time as Britain's official standard time should have helped matters. But outside the big cities, old practices took longer to change. In 1901, a further Factory Act showed that nothing had really moved on since 1844:

> Where an inspector, by notice in writing, names a public clock, or some other clock open to public view, for the purpose of regulating the period of employment in a factory or workshop, the period of employment and the times allowed for meals in that factory or workshop shall be regulated by that clock.[5]

In the factories and mills of Manchester and the northwest, time at the turn of the twentieth century remained a free-for-all, and it was not just the workers who felt they were losing out. Factory managers were being prosecuted repeatedly by the factory inspectors for illegal overtime because all the official clocks told different times. By 1913, the Oldham Master Cotton Spinners' Association had had enough, describing its members as "the victims of injustice" when these prosecutions took place.[6] It is fair to say that this probably did not create an outpouring of sympathy, but the prosecutions nevertheless prompted the association to put forward a solution that would have looked familiar to any progressive city across Europe, from London to Brno. It wanted to build a network of electric clocks across Lancashire, all telling the time of a single driving-clock in the association's office, set to Greenwich Mean Time by an electrical signal from the observatory.

———•———

STANDARD TIME SHOWN on public clocks is an instrument of moral control, and it is inextricably bound up with electricity, because time signals need to travel everywhere in an instant. Without electricity, a cotton mill in Lancashire, several miles from the nearest telephone exchange or post office, could run without Greenwich time for decades after the railways supposedly standardized Britain to GMT. Even London pubs could run on their own time until the lawmakers and engineers gave temperance campaigners the tools that they needed to restrict the sale of alcohol. Electricity, carried on overhead wires, could do more than standardize time. It brought these isolated islands of time, these lawless enclaves of temporal disorder, whether in Oldham or Old Street, into the hands of the moral reformers and their clocks. Once the system had been proved, in Britain and in modern cities around the world, there was no limit to the moral engineering it could achieve.

———•———

WHEN I WAS not working my way through London's historic pubs with James Nye, I spent a lot of 2007 retracing the steps of William Willett, a property developer born in 1856 who was responsible for large housing estates across London and Brighton. Willett was obsessed by daylight. He had joined his father's building firm when he was in his early twenties and they spent the next quarter century building what commentators still consider a higher class of housing, with plots carefully laid out to maximize light entering the large and plentiful windows of the houses.

But Willett's obsession went far beyond business: he believed in the moral value of daylight with what amounted to messianic zeal, and his conversion, if not on the road to Damascus, at least on the bridle path in the woods near his Chislehurst home, led him to the project that made him famous to this day. His daughter Gertrude recalled the scene: "Every morning before breakfast he rode through Petts Wood . . . There were beautiful bridle-paths leading through the woods under the pine trees; it was here that he first thought of Daylight Saving."[7] The epiphany occurred in about 1906, and Willett spent the rest of his life campaigning for legislation to change the time on all our clocks and watches every spring and autumn to make better use of summer daylight.

Daylight Saving Time, known in the UK as British Summer Time, relabels the hours on every clock and watch for a few months each summer; 9 a.m. becomes 10 a.m., for instance, so if we must be at work for nine by the clock, we must get up an hour earlier. It is that simple. There is no extra daylight (of course), we are simply getting up and going to bed an hour earlier in summer than in winter, which makes the summer evenings feel lighter (because we are going to bed earlier), and we are doing it because William Willett believed it would be good for us.

The moralists came out in force to support Willett's clock-changing scheme when he began to promote it in 1907. Arthur Conan Doyle heaped praise on the plan by claiming that "it is a splendid thing for every man in summer to get back to his home in time to look after his little garden, or whatever his particular hobby might be, after his day's work."[8] John Lubbock, Lord Avebury, the banker who gave us bank holidays, claimed Willett's proposals would help London's huge population of clerks, who "would be able to get away at a time which would enable them to play a game of cricket or get some other healthy outdoor exercise, which I believe would be a very great boon to them."[9] In 1911, the Home Secretary, Winston Churchill, claimed that "a grateful posterity, dwelling in a brighter and healthier world, would raise statues in honour of Mr Willett and decorate them with sunflowers on the longest day of the year."[10]

Willett did not live to see his scheme put into action. He campaigned vociferously and extensively for his clock-shifting scheme until his death in 1915, and managed to get two Select Committees to consider the matter in Parliament, but it was only in 1916, during Britain's war with Germany, that the German military figured out that daylight saving might be a good way to save fuel used for lighting their munitions factories in the evenings, so they tried it out, followed by similar trials in Austria-Hungary, Holland, Belgium, Denmark and Sweden. A few weeks later, not to be outdone, the UK followed suit, and the scheme grew from there.

In the early days, we knew we were being hoodwinked by this moralizing housebuilder from Chislehurst. On the first day of Britain's 1916 summer-time trial, a newspaper journalist spotted graffiti chalked on the pavement near Blackfriars station which said bitterly, "All Fools' Day, May 21. Get up one hour earlier and kid yourself you haven't."[11] But, still, we went along with the lie, because it had powerful backers.

From the moment they were first put forward, the electric time industry had seen William Willett's moralistic proposals as a Trojan horse to get its technology further embedded in the official fabric of urban life, and it made the most of the 1908 Select Committee hearing that considered the scheme.

Picture the scene on Tuesday, May 26, 1908. In the committee chair was Edward Sassoon, MP for Hythe and married into the Rothschild banking dynasty, with a keen interest in the state control of electric telegraph networks overseas. In front of him came St. Andrew St. John Winne, chair of the Standard Time Company, which was still selling Greenwich time to London's pubs, offices and banks (including Rothschild's). Winne explained that Willett's scheme would be easy to put into practice using his company's electrical synchronization equipment, even setting up a technical demonstration on the committee room table to show it in action. The next witness was Frank Hope-Jones, founder of the Synchronome company, which made electric time systems exactly like the ones installed in Brno a couple of years earlier, and the ones which the Oldham Cotton Spinners' Association would want to set up across Lancashire five years later. Hope-Jones spoke passionately to the committee about the new science of electric time systems, describing them as "a saner method of timekeeping," and warned that the UK's methods of standardizing time were "very far behind those which are commonly used by almost every other civilised country."[12]

Of course, William Willett was not the first moralist to rail against idleness in the working classes, nor the first to try using clocks to cure the problem. In a sense, his position was the result of centuries of moralizing about the importance of clocks in creating discipline, whether in the service of religion or of industry. But electricity gave clocks new powers, immense reach and an elevated moral status in society.

Electricity *is* modernity. As the nineteenth century turned into the twentieth, electricity was changing every facet of everyday life: how we communicated, how we moved around, our health, our well-being. It changed how we saw our world and ourselves. The electrical future filled the pages of novels and magazines; it was vividly portrayed in art and advertising. It was romantic, thrilling—shocking, even. It is hard to overstate how convulsive were the changes to our lives with the development of electricity and its application to ancient technologies like clocks as well as more recent ones. The moral makeup of Western society was being reshaped by this new modern force.

In the Select Committee meeting room considering Daylight Saving in 1908, the electric-clock promoters were selling modernity, as they always had. This was just three years after Albert Einstein had published his theory of special relativity, with its mind-bending proposal that the speed of light was absolute, but time was relative depending on your viewpoint. The STC, the Synchronome company and Johann Antel in Brno were all offering standard time at the speed of light, time that could be changed depending on the moral standpoint of the politicians in power at the time. Alcohol, child labor, sloth: whatever the moral crisis, electricity and clocks could put it all right.

Time that has been standardized allows those in power to control the behavior of everyone in the standard time zone—when you wake up, when you go to bed, when you can and cannot buy alcohol, how long you are allowed to work in factories and shops, how much daylight you get in summer when the days are longer, how easily you can travel to other

places. The people in power are deciding what is good or bad behavior—
what is morally right or wrong. And they are using clocks and electricity
to enact their moral code. Each time you fight your way to work on a
crowded bus, or sit in traffic during rush hour, you are obeying electric
clocks; more to the point, you are obeying the commands of the government, which uses time to tell you how you should behave. For centuries,
time has been used to keep populations in line. With the new electrical
systems, moral control could be applied at the speed of light. And we
always obey. Or do we?

Resistance

Telescope Driving-Clock, Edinburgh, 1913

As the local evening papers hit the newsstands of Edinburgh on Friday, May 16, 1913, the talk on residents' lips was the terror campaign being waged by suffragettes, who that morning had left a bomb in a boiler room underneath a chapel in nearby Dalkeith, owned by the Duke of Buccleuch. The device was discovered by an estate worker before it could be detonated, but it was clear that it would have killed the entire congregation of the church above if the plot had succeeded. Readers also discovered that the chapel contained the burial vaults of the Buccleuch family, which would only have added gruesome indignity if the bomb had been set off. Some commentators wanted the army to be deployed to help guard property, claiming that Scotland Yard was overwhelmed by the suffragette activities sweeping Britain at the time. It seemed that nowhere was sacred.

That night, John Storey, an assistant at the Royal Observatory, located high on Blackford Hill to the southwest of Edinburgh's city center, spotted a suspicious figure loitering outside the observatory buildings. He raised the alarm with his boss, Ralph Sampson, Scotland's Astronomer

Royal, an Edinburgh resident who had more reason than most to fear the suffragette threat. Sampson's observatory was filled with valuable and delicate astronomical instruments used not just to study the stars but to keep the most accurate time, and it was run only by himself and a couple of assistants. He knew how vulnerable they were to attack. The next day, Sampson wrote to the chief constable of the City Police asking for help in guarding the observatory, only to be told that he would have to cover the costs out of his own staff budget. A few days passed as Sampson mulled over the problem. But he had been right to suspect an attack. Before he could put any additional security arrangements in place, Sampson's worst fears were realized. A bomb explosion ripped through his observatory.

The weather in Edinburgh on Wednesday, May 21, was blustery and wet, so when Sampson was woken up in the early hours of the morning by a loud noise, he thought it was just a door banging and went back to sleep. But a large earthenware jar, packed with gunpowder, had been placed in the middle story of the western telescope dome, next to an iron staircase, connected by a thirty-foot fuse that had been run down the staircase to the room below, which housed the telescope chronograph, a clockwork device for timing observations. And when his servant started to open the telescope buildings the next day, the wreckage was revealed.

The damage caused by the explosion was extensive. Windows and doors had been blown out. Plaster ceilings were shattered. Some of the staircase was blown off. Splinters of glass were recovered 100 feet from the building. The heavy floor of the telescope room directly above the bomb had been knocked off the corbels that supported it, and had been badly damaged, although it did not collapse, and had shielded the two fragile telescopes housed in the dome. The driving-clock, which kept the telescope fixed on a star while the Earth rotated, came off worse, although it was later repaired.

There was little doubt which group had carried out the bombing that night, although the individuals involved were never identified or

Officials investigating bomb damage at the Edinburgh observatory, 1913

caught. Two notes were left at the scene. The first read, "From the beginning of the world every stage of human progress has been from scaffold to scaffold and from stake to stake." The other said, "How beggarly appears argument before defiant deed. Votes for women."[1] This bombing, described in the press as an "outrage" and "sensational," was the work of suffragettes campaigning for equal suffrage.[2]

WHILE EARLY CAMPAIGNS for women's suffrage were anti-violence, by the 1910s the emerging suffragette movement was becoming renowned for its violent protests across Britain, with targets including racecourses, parks, tourist attractions and men's clubs. Communication and transport systems were also frequently hit, with acid being poured into mailboxes, arson attacks on railway carriages, and the cutting of telephone lines and fire alarm circuits.

Why would an astronomical observatory, with its time-finding tele-scopes and clocks, be considered a suitable target for a bomb? The answer comes from once again considering the standardization of time. Standardization was an act of power and control. Science itself, or rather scientific activity, was the enaction of power in many senses. And the global systems of transport and communications that relied on standard time allowed power to flow around the world.

This was a process that had been growing in scale and effect since the 1830s, when the introduction of railways and electric telegraphy had created the demand for a single, standard time across a large geographi-cal area, as well as the technological means by which it could be shared. Railway timetables worked well only if there was one "railway" time across the network. But what started as a convenient industrial coordina-tion technique grew into a system that saw standard time embedded in the entire fabric of Victorian and Edwardian life. As standard time swept across the globe, clocks went hand in hand with trains and telegraphs in an international system of commerce, trade and empire—a system of power. All this power was wielded by men. The beneficiaries of glo-balization were men. The scientific standardization of time took place in a man's world. No wonder Ralph Sampson had feared his scientific observatory was at risk. His clocks and telescopes were a clear target for attack by suffragettes, who felt the system was stacked against women. For them, Sampson's instruments looked like tyrants.

The Canadian anarchist writer George Woodcock spent the Second World War as a conscientious objector working as a farm laborer in Essex. In 1944, while there, he wrote an essay called "The Tyranny of the Clock" in which he claimed that "the clock represents an ele-ment of mechanical tyranny in the lives of modern men more potent than any individual exploiter or any other machine."[3] The idea of clocks as tyrants has stuck with us ever since, as clocks have been made by tyrants to enact their control and coercion in all the myriad ways we have explored thus far.

But the thing with tyrants is that people try to resist them. Sometimes this is through loud, singular, violent acts: assassinations or bloody coups. At other times resistance is mounted by long-drawn-out accumulations of multiple small, quiet acts which, together, can build into a powerful and sometimes unstoppable force. It does not always work, at least not in the short run, and of course one person's tyrant is another's freedom fighter. But history has shown us, again and again, that tyranny is rewarded with resistance. So, if you want to understand civilizations and the resistance to tyranny, see how subjected people treat clocks. Wherever you look, you will find acts of resistance against them.

———·———

IN 1967, WRITING as the Summer of Love was reaching its climax in San Francisco and hippies in London were following the lead of their American counterparts, the Marxist historian and social activist E. P. Thompson drew an analogy between the dropout culture he was witnessing and historical forms of resistance against clocks and time discipline. He commented that "One recurrent form of revolt within Western industrial capitalism, whether bohemian or beatnik, has often taken the form of flouting the urgency of respectable time-values."[4]

Time historians have been arguing about this ever since. On Thompson's side are those who believe that rising industrialization from the eighteenth century onward caused a *new* form of time discipline over working-class people, who were increasingly paid by the clock rather than the job. But many (probably most, now) refute his claim that the rise of industrial capitalism forced workers into new forms of bondage mediated by clocks, seeing instead the countless ways workers have been disciplined by the clock as far back as we care to look, and in places besides Britain or the industrial West.

But whatever the reasons, wherever and whenever it took place, the use of clocks to control workforces and populations has cast a long shadow, and sometimes people have fought back. The UK's industrial factories and mills certainly had a problem with time: clocks were part of the regulation of factory labor, leaving the system wide open to abuse. As late as the nineteenth century, a Dundee textile factory worker was able to claim that:

> The clocks at the factories were often put forward in the morning and back at night, and instead of being instruments for the measurement of time, they were used as cloaks for cheatery and oppression . . . it was no uncommon event to dismiss any one who presumed to know too much about the science of horology.[5]

The textile industry was one of the most oppressive in its use of clocks to discipline the lives of its workers, and the most blatant in the way its owners and managers cheated their workforce out of time, and therefore money. Knowledge is power, and workers in the earliest factories and mills were forbidden from carrying watches, for fear they would know how long they were working beyond their paid hours. The time on the factory clock was routinely changed over the course of a day.

But some managers went further by collaborating with clockmakers to construct ever more ingenious mechanisms and technical modifications to cheat workers out of the true time and lengthen their days. In 1832, the British Parliament was debating the so-called Ten Hours Bill on textile factory reform. Members were shocked to be told by Michael Thomas Sadler, MP for Newark, social reformer and promoter of the Bill, that:

> a practice is known to exist in certain mills or factories of using two or more different clocks or timepieces, one being a common or Time Clock, and the other a clock regulated by the velocity of the steam engine or other machinery, and often called a Speed Clock, by which the daily labour, though nominally limited to a certain duration, is often increased much beyond that limitation.[6]

A few of these speed or engine clocks have survived—certainly enough to know they were quite common across the textile industry. One, dating from 1810, was used at Park Green silk mill in Macclesfield, Cheshire, with a dial and hands connected by shafts and belts to the waterwheel powering the mill. If the river was low, workers had to work longer. Another was made by John Barrett of Skipton, North Yorkshire, calibrated in terms of the factory's working shift, not the twenty-four-hour day. A bell signaled the passing of each half hour of "engine time." But an earlier survival, an engine clock made by the renowned clockmaker and scientist John White-hurst in about 1770 for Josiah Wedgwood's pottery factory in Stoke-on-Trent, Staffordshire, is more noteworthy, as Whitehurst and Wedgwood were both friends of Benjamin Franklin, the American revolutionary politician who had first equated time with money in 1748. Where money was involved, time became relative. The clock's role as oppressor or as savior of the oppressed depended on who *controlled* the time it told.

So, if the managers of the industrial revolution's factories and mills had weaponized clocks and watches, it must come as no surprise that some people chose to resist. Often this was through absenteeism but sometimes it took a sharper form. The term "Luddite" has entered our language following the activities of a group of machine breakers in the early nineteenth century, who destroyed looms and steam engines. The Luddites themselves, rioting from 1811 to 1813, met with harsh justice. Many were executed. But resistance to the tyranny of machinery cropped up frequently, such was the powerful symbolism of technology. In 1826, thousands of armed rioters marched into the Lancashire cotton town of Accrington, heading straight for the Sykes mill, where, one witness reported, "The first thing was that of a woman smashing a clock that hung in the passage. The next was an onslaught on the looms with crowbars and sledge-hammers."[7] As late as 1878, factory clocks were being destroyed or stolen by attackers at several mills in the Lancashire textile town of Blackburn. Cotton workers there considered themselves "slaves" to the factory bell.[8]

Over in the cotton, tobacco and rice plantations in the pre–Civil War American South, plantation owners were closely watching developments in the industrial North of America, and in Britain. They saw the powerful disciplinary effects of clock time as it became established in the mills and factories, and they wanted to gain membership of this rapidly growing capitalist club by importing it into the fields of the South. As the historian Mark Smith has described, "the discipline provided by the lash would be rendered of secondary importance to one governed by the clock."[9]

It was harder, in some ways, to import industry's time discipline into a world where workers were enslaved for life, but clocks (and their representatives in the fields, namely watches, bells and horns) did begin to exert their control over enslaved workers. As with the Lancashire mills, they met with resistance. One South Carolina planter in the 1830s said that "Negroes thus employed, that is, working by time, it is well known move much more tardily than when tasked," and a Mississippi planter a decade later complained that his enslaved workers refused to eat as fast as the whites on the plantation. Servants might wake plantation visitors early, or late, or cook their meals at the wrong times (according to the house clock).[10] There was resistance, too, in southern Africa. As Keletso Atkins has shown, South Africa's black workforce repeatedly took direct action over the use of time in their working lives under white European control. Even as late as the turn of the twentieth century, hundreds of dock laborers in Durban struck—successfully—for overtime payments that had been denied them. If the whites were trying to embed time discipline into the very lives of black people through constant repetition, then the acts of resistance against clock time were just as powerful and repetitive.

By the 1870s, as clocks in workplaces around the world were beginning to tell times that were standardized and centralized in scientific observatories, a new political force was on the rise that offered a fresh theoretical focus on the structures of Western time. The Russian revolutionary anarchist and founder of social anarchy, Mikhail Bakunin, argued in the 1870s that while science might be legitimate, it could not be universal. It could not go outside abstractions. It could not grasp the concrete aspects of life. Bakunin preached what he called "the revolt of life against science, or rather against the government of science."[11]

In an 1871 essay, Bakunin talked about a learned academy of scientists. It was a fictional academy, but you could imagine it comprised men of science like those running the observatories at Greenwich and Edinburgh at the time. Bakunin claimed that society run by an academy such as this:

> would be a monstrosity . . . for two reasons: first, that human science is always and necessarily imperfect . . . The second reason is this: a society which should obey legislation emanating from a scientific academy . . . would be a society, not of men, but of brutes . . . It would surely and rapidly descend to the lowest stage of idiocy.[12]

He described how he refused to recognize a single, universal authority:

> if such universality could ever be realised in a single man, and if he wished to take advantage thereof to impose his authority upon us, it would be necessary to drive this man out of society, because his authority would inevitably reduce all the others to slavery and imbecility.[13]

That word "universal," and that notion of monstrous scientists "imposing their authority upon us," is telling in this context of universal time, with time standardization becoming important in the 1870s when Bakunin wrote his essay and reaching its ultimate political conclusion in 1884. In November of that year, the International Meridian Conference held at Washington, DC, came to an end, resolving that Greenwich should be chosen as the prime meridian of the world, measuring all time and space from the meridian defined by the great transit telescope in the Greenwich observatory. What was being discussed when Bakunin published his theories on the role of science in society was a single, universal time for the entire world.

As we have seen, astronomical observatories, far from being temples of abstract science, have always been intensely political domains.

The standardization of time and space is all about maintaining order, controlling people's behavior, making money, securing political power, building empires and waging war. It is authoritarian and it is nationalistic. The suffragette bombing of the Edinburgh observatory was not the first time a royal observatory had been the target of violent attack. In Greenwich, in 1894, the anarchists had got there first.

THE LAST TWO decades of the nineteenth century saw a series of anarchist-inspired terrorist attacks hit countries across Europe. One of the earliest and most spectacular was the bomb assassination of the Russian tsar, Alexander II, in 1881. This led anarchists to carry out numerous similar attacks on rulers and the aristocracy. At the same time, Irish nationalists resisting British colonialism began a dynamite bombing campaign on the British mainland.

In January 1885, just a few weeks after the Washington Meridian Conference shone a spotlight on his observatory, the Astronomer Royal at Greenwich, William Christie, wrote to the government, worried about possible attacks on his own buildings. Noting "the repeated attempts to blow up public buildings with dynamite," he claimed that "it is conceivable that serious damage might be done from the outside, which is accessible to the public frequenting Greenwich Park."[14] Nothing came of his request—a bureaucratic brush-off—and Christie let the matter drop at that stage, but anarchist attacks continued across Europe.

Just three weeks later, with tensions rising across London, an anonymous letter containing threats to blow up St. Paul's Cathedral prompted authorities to close off the cathedral's clock and bell chamber, high up in its southwest tower, fearing a symbolic act of violence against the clock. It remained closed for years.

By late 1893, anarchist terrorists were particularly active in France, culminating in a bombing of the Chamber of Deputies in Paris that December. On January 4, 1894, one of the astronomers at the Greenwich observatory reported a break-in to buildings under construction at the south of the site. The next day, William Christie wrote again to the government. "As this portion of the Observatory premises . . . is much exposed at this season of the year to attack by dishonest or mischievous persons . . . I would submit that immediate steps should be taken to provide proper police protection to supplement existing arrangements."[15] He talked about how, when the park was closed to the public after six o'clock in the evening, there were no regular patrols by either park keepers or police. But he was more worried about the period just before closing time and demanded increased park patrols and the provision of police protection. The government sent an acknowledgment but nothing more.

Three weeks later, on the evening of Monday, February 12, 1894, an anarchist terrorist threw a bomb into the crowded café of the Grand Hotel Terminus in the Rue Saint-Lazare, Paris, next to a major railway station. The café was packed with customers listening to an orchestra when the anarchist threw in a small box made of a sardine case filled with explosive and lead. Twenty people were injured in the explosion, and one of them later died from his injuries. The bomber ran out, was chased, fired five shots with his revolver and stabbed pursuers with a knife while trying to escape, wounding several more people, before he was captured and arrested. He had been living in London and said that his aim was "to give a warning to the bourgeois Government which is so hard on the poor and miserable."[16] As it turned out, police later believed the bomber's real target was the Comédie Française nearby, which was showing the first performance of a major new play and was full of the elite of Parisian society. Too full, in fact, for the terrorist to get in, so he took his bomb to the café. An explosion in the packed Comédie Française would have been too horrific for words.

Three days later, an explosion was heard across Greenwich. The Royal Observatory gate porter and two of its astronomers raced out of the observatory complex toward a path where they could see smoke rising. There they found the park keeper, Patrick Sullivan, and two pupils from a local boys' school who had been first on the scene. At first sight, little

RESISTANCE

seemed wrong. A man, later identified as Martial Bourdin, was kneeling
on the path by railings, perfectly still, with his head bowed. Then he
fell forward. Sullivan gently lifted him upright and asked, "What has
happened?"[17] But Bourdin did not answer, and for the first time, the
gathered party could see the horrific extent of his injuries. His left hand
had been blown off above the wrist, with bloody sinews and tendons
hanging down. His shoulder blade was protruding through a hole in his
back. And the explosion had blown his intestines out. He was taken to
a nearby hospital where he died twenty-five minutes later. He never said
what had happened.

The government quickly sent investigators to the scene of the explosion, where they found the path heavily bloodstained and littered with
fragments of flesh and bone from Bourdin's hand. But, apart from damage to the face of one brick, the observatory itself had not been harmed.

A terrorist act against the Royal Observatory, resulting in a gruesome
and public death, meant that the bomb attack hit the headlines with
deadly force for days afterward. All the grisly details of Bourdin's injuries
were quickly released, and hordes of tourists flocked to Greenwich Park to
view the scene, where they could inspect white stakes planted in the grass
that marked the sites where parts of his hand had been found. One keeper
said there were more visitors to Greenwich Park then than ever before.

In 2008, when I worked at the Royal Observatory as its curator of
timekeeping, I was studying the history of the explosion, and the years
of political analysis that had followed. But something was missing.
Throughout the accounts I read, I kept finding doubts being expressed
about Bourdin's intentions that day. Some said he was passing through
Greenwich Park in search of a busier target, or on his way to Dover
to flee the country in the face of increased police surveillance. Others
believed he was looking for a secluded spot to hide explosives. But none
of this added up to me. In theory, all these explanations make sense. But
Greenwich Park is a real place, not a theoretical one. It has hills and hol-
lows; trees and bushes; pathways and sightlines, and that specific topog-
raphy had largely been ignored by those writing about the incident in the

years that followed. So in order to get a fresh understanding of the events of that fateful day, I decided to retrace Martial Bourdin's final journey—and continue it to its planned conclusion.

Studying an Ordnance Survey map of the park dating back to the 1890s, I could clearly see drawn on it the zigzag path that Bourdin had taken up to the observatory, which sits high on a hill. I had seen the government sketch plan of the scene as well as photographs and engravings in the press that showed a steep hillside with a few trees dotted here and there, with the path threading its way up toward the northwestern turret of the observatory's main building.

Government sketch showing the scene of the Greenwich observatory explosion, 1894

Today, the path is long gone. The area is entirely overgrown with trees and scrub and has been fenced off from the public, though my observatory colleague Jonathan Betts and I had special permission from the park authorities to enter. A more recent path now takes visitors on a different route up to the observatory and beyond it toward Blackheath. I think this later path is what commentators have assumed was facing Bourdin. But the actual path he took, which we carefully retraced by forcing our way through the undergrowth, led only to the observatory. Nobody would have gone that way who simply wanted to pass through the park, even if they were unfamiliar with its layout. It could be read, from a distance, at a glance. And the path was not secluded in those days, either. Anybody climbing it was clearly visible to passersby. It was hardly a place to hide explosives.

As we fought our way up the hill, we reached the spot where Bourdin had been fatally injured. The bomb was in his hand and had been primed, ready to go off. But he was not a suicide bomber. Somehow, possibly on a tree root or an uneven spot on the path, Bourdin had tripped and fallen forward, setting off the bomb by mistake and landing on it as it exploded. The task that faced us now was to continue the journey he would have taken—the *only* journey available to him—to see if we could work out what his target had really been. Continuing up to the top of the hill, we reached the path that hugs the side of the observatory complex itself and followed it around. To our left, the open hillside fell steeply away. To our right were the high brick walls of the observatory. A small bomb, like the one Bourdin was carrying, would never have made much of a mark here, and I could not believe that was his target.

Then, as we rounded the corner of the observatory complex, we reached the gates that lead into its large open courtyard. And there, beside the gates, is the large, round, white dial of the observatory's public clock, set in a wall at head height, behind glass. It is an official clock that has displayed Britain's standardized, centralized time since the nineteenth century. In 1884, the Western world's government representatives had decided that all the people of Earth should march to the beat of

Greenwich observatory official posing with
gate clock, c. 1925

one clock—and it was *this* clock. It was the powerful, living embodiment
of all that anarchists like Bourdin despised, and it must have looked like
a monster. I realized that I had reached the end of our journey. There was
no other target. Martial Bourdin had been moments away from bombing
Greenwich Mean Time itself.

———

THE 1884 MERIDIAN Conference that selected Greenwich for the world's
prime meridian opened what was to become a running sore in the psy-
che of the French nation. At the time, the French delegates lobbied hard

for the decision to be reversed, and later simply refused to recognize the results of the conference. French commentators saw it as an act of imperialism on the part of the United Kingdom, and it is easy to see their point. Whatever the reason in practice for choosing Greenwich—it was convenient because most ships around the world used charts based on the Greenwich meridian—in principle it looked to some people like Britain attempting to impose its values on the whole world. It looked like a hostile act of colonialism.

The conference in 1884 did not result in immediate action, such is the nature of global diplomacy. It took until the early 1890s for a majority of countries to shift to a Greenwich-based time system, which meant that it stayed in the news, drip-feeding hot bile into the French people and fanning the flames of resentment, for years. Then, in late 1893, a series of surveys took place to work out who in the world was keeping time based on Greenwich, which led to a variety of popular articles appearing that talked (again) about this inexorable rise of Greenwich-based universal time.

So, by early 1894, far from the Meridian Conference being ten-year-old history, in fact it was very much in the news headlines and still a current and contentious story. The connection between the Greenwich observatory and wounded French pride was made at the time of the 1894 bombing in the national press. *The Times* stated that "the fact that the reputation of Greenwich Observatory is world-wide, and that Frenchmen have rather an objection to its pre-eminence, may have been influential in the mind of a Frenchman who was clearly . . . fairly educated."[18]

Émile Henry, the anarchist who threw the bomb into the Hotel Terminus in February 1894, three days before Bourdin's bomb, said at his trial:

> You have hanged us in Chicago, decapitated us in Germany, garrotted us in Xerez, shot us in Barcelona, guillotined us in Montbrison and in Paris, but what you can never destroy is anarchy. Its roots are too deep, born in a poisonous society which is falling apart; anarchism is a violent reaction against the established order.[19]

Henry knew Bourdin well, and his choice of words there is telling. "Established order" means political institutions and structures, but it also refers to the work of measurement scientists. Their work is in establishing order. Then, as now, it was impossible to separate science from society, to see the work of any institution as purely scientific rather than bound up in politics and wider culture.

The role of the Greenwich observatory in the Meridian Conference, and the status of GMT as the defining time for the world, was intensely concerning to people holding anarchist views, especially anarchists based in France. A blow against the observatory would have been a blow at the prime meridian and GMT and therefore a blow against state-sponsored scientific imperialism. Anarchists believed that authority was the root of all exploitation, and only an anti-authoritarian and decentralized society could ensure human freedom.

·   And the anarchists of 1890s France were not the only people resisting Britain's colonialism by attacking its imperial clocks.

———

IN 1898, FOUR years after the Greenwich bombing, the Indian city of Bombay (today's Mumbai) was gripped by rioting and widespread strikes by the native Hindu and Muslim population resisting harsh and repressive public-health measures being carried out by the local British government. At the geographical, social and strategic heart of the city, dividing the native and European areas, sat the British-built buildings and stalls of the Crawford Market. Rising above it, overlooking a busy intersection and facing Bombay's police station, was the market's clock tower. On the evening of Friday, March 11, 1898, as anti-colonial unrest grew across the city, native Indian attackers with rifles opened fire on the clock, partially destroying one of its brightly illuminated dials.

A long-running dispute over time standardization in Bombay in the 1880s had meant that public clocks had taken on a particular symbolic meaning in the city: the British had tried to force their standard time

Crawford Market and clock tower, Bombay, in an early-twentieth-century postcard

(actually Madras Time) on to Bombay. It had been resisted, but the argument was still fresh in the memories of the Bombay population in 1898, when the Crawford Market clock was attacked. Seven years after that, in 1905, arguments over standard time in the city re-emerged, as the British government tried to unify all of India's population under a single time zone, five and a half hours ahead of Greenwich.

At a time of intense political tension over British colonialism in India, standard time in Bombay took on ever more powerful symbolic meaning. The choice of time kept on the public clocks across the city—Bombay Time, Madras Time or the new Indian Standard Time—was an expression of colonial allegiance—or resistance. After weeks of increasingly bitter argument, as the new time standard was characterized as an unwelcome European colonial imposition on the native population of Bombay, the discussion that had previously been held behind closed doors spilled over. In December 1905, mass public demonstrations broke out, with one gathering attracting 15,000 protestors. The following month, the largest textile mill in Bombay changed its clocks to the new standard time without informing its 4,500 workers. When they turned up that morning to begin their shifts and discovered what had happened, they immediately went on strike and began to pelt the mill's clock tower with stones.

Protests about Bombay's time were really an expression of unrest among the native populace about British colonial rule. Clocks could be weaponized by local politicians who wanted to affect public opinion on wider political issues. As agents of centralized control, power and domination, clocks arouse and inflame human emotion, which can lead to resistance—to violent attack.

Authority breeds resistance. Standardization gives rise to dissent. People fight clocks, and they have always done so. Because what we are really doing is fighting with each other, as we have poured our very identities into clocks.

Identity

Golden Telephone Handsets, London, 1935

Usually, when she left her house in Jarrow to head off for work, Mary Dixon made her way to the nearby town of South Shields, at the mouth of the River Tyne where it meets the cold, gray North Sea. She was a supervisor at the town's telephone exchange, spending her working life with the operators who connected telephone calls and answered requests for the time—100,000 time requests every month just in London alone. But today, instead of heading for South Shields, Dixon made the longer journey into Newcastle city center, getting out at the central railway station before climbing on board a fast steam train for the five-hour journey to London. Dixon was excited but nervous as she set off for the capital, because this was not just an ordinary working week. Tomorrow might be the day she was propelled on her way to national fame, and she had eight competitors to fight off armed only with her voice. It was June 1935, and she was on her way to the grand final of the competition to find the voice of the telephone speaking clock.

The General Post Office, which ran the UK's telephone network, had been hosting the competition to find what they called the "Girl with the

Golden Voice" since April. A series of regional heats had seen 15,000 auditioning exchange operators whittled down to the nine finalists who were gathering in London. Mary Dixon had already beaten rivals in tests held at Newcastle and then Leeds. Could she score the hat trick at the grand final?

It was a nerve-wracking experience for the nine women who gathered at the GPO building near St. Paul's Cathedral, taking their places in turn at a microphone in a private room, connected to golden telephone handsets placed in front of each member of the judging panel that was seated around a big table in the building's main hall. Some of the finalists were local. Four of the women worked in London exchanges and a fifth came from nearby Guildford. But others, like Dixon, had traveled long distances to represent their districts: Birmingham, Exeter and Blackpool, as well as South Shields.

Dixon was one of the older finalists. Telephone exchange operators usually had to leave their jobs when they got married, so most operators were in their late teens or twenties. Dixon, who never married, was able to rise through the ranks to senior operator and supervisor, and she was forty-two years old when the golden voice competition entered her life.

A lot rested on this. Regional pride, as well as the chance of lucrative promotional opportunities, was at stake. Gathered round the judging table was the cream of Britain's acting, literary and business communities. Sybil Thorndike, the world-famous Shakespearean actor, commanded the room. The BBC's chief announcer, Stuart Hibberd, sat alongside, ready to pass judgment over the women's diction. The newspaper baron Langton Iliffe represented the world of commerce, and Rita Atkinson had come all the way from West Yorkshire, chosen in a parallel competition to find the "perfect telephone subscriber."

The final judge was the UK's poet laureate, John Masefield, who had been asked to set the speaking tests. The winning voice, he had insisted, needed to be impersonal. "It is to come like the voice of a nightingale singing in the midnight," he said, "without any trace of over emphasis or personal advertisement." It was to be beautiful in every respect. "Any trace of the theatrical would be fatal."[1] Masefield originally wanted to set the finalists readings from the Bible's Book of Ezekiel and English translations of the Greek writer Aesop's *Fables*, but later settled on verses from the seventeenth-century poet John Milton's "L'Allegro" and passages from Robert Louis Stevenson's *Treasure Island*, as well as more prosaic time sentences. The judges listened intently on their golden handsets as the finalists spoke their lines, scribbling down marks on special score sheets for each candidate's purity of tone, clarity, accuracy, "pleasant inflection" and "freedom from accent or other voice peculiarity."[2] In the end, Ethel Cain, an operator at London's Victoria telephone exchange, won the contest.

After the decision had been made, John Masefield came bounding over to the journalists waiting at the side of the hall and told them how happy he was with the result. "Miss Cain has one of the most beautiful voices I have ever heard, and behind a beautiful voice you will also find intelligence," he beamed. "We tried to get a voice that was no more personal than a bird in a bush: I think we have succeeded."[3] Ethel Cain's success meant the end of a dream for Mary Dixon. Masefield commented that out of all the finalists, "only in one were we able to find any trace of accent, and that was a slight Northern one."[4] Dixon's Geordie heritage had cost her the chance to be a star, and she returned home, dejected, to Jarrow, where she was comforted by her sisters before returning to work at the South Shields telephone exchange.

The ceremonial switching on of the speaking clock was a moment of great pomp. It was July 24, 1936, and seated in a reception room of the Holborn Telephone Exchange, London, were the great and good from horology, telecommunications and local politics, as well as Ethel Cain herself and Rita Atkinson, reprising her role as the perfect subscriber. The postmaster general, George Tryon, made the first speech:

> Men have sought to reckon the passage of time by water flowing
> from a vessel or by sand in an hour glass, or by the sun-dial. We
> have clocks that convey information by sound, the alarm clock, and

Ethel Cain, photographed after winning the
"Golden Voice" competition final, 1935

the striking clock. Here we have the latest and most wonderful of
clocks. It only speaks when spoken to. It is accurate to within one-
tenth of a second and it tells you the time in the pleasing tones of
what is known as the "Golden Voice."[5]

Then the Astronomer Royal, Harold Spencer Jones, spoke about the time-
measurement activities carried out for the nation at the Royal Observatory, Greenwich, which was providing the time for the new telephone
clock. After he had finished, Bertram Cohen, the engineer responsible
for building it, gave a technical presentation.

Finally, it was the moment of truth. Spencer Jones lifted the receiver
of a telephone handset and dialed "TIM," the first official call to the new
speaking clock. A loudspeaker reproduced the sound of the clock to the

assembled throng. Then the mayoress of Holborn, Katherine Langdon, made her own call to TIM. Once the switching-on ceremony was over, tea was served, and everybody went down to the basement to see the clocks themselves, as their glass discs containing Ethel Cain's golden voice rotated slowly inside their glazed cabinets, dispensing Greenwich time at the third stroke, precisely.

It was a momentous day for clocks in the UK, and for voices. It was also rather a sad day for my family.

I AM DISTANTLY related to Mary Dixon. Very distantly. But my mother knew her well. Dixon was a kind woman, appreciated in the family for having a wonderful sense of humor, although I have been told that she could appear rather serious to those who did not know her. A family photograph bears this out.

Family photograph of Mary Dixon (right) with her older sisters Anne (middle) and Margaret (left) outside their home in Jarrow, 1930s

It was exciting to learn that somebody in my family was *almost* the most listened to voice in 1930s Britain. This was one reason why I wanted to explore Dixon's experience of the golden-voice competition as well as that of Ethel Cain, who ended up winning the contest and going on to fame, if not fortune.

But I had another reason for telling Dixon's story. Even today, in the third decade of the twenty-first century, the UK speaking clock remains a firm fixture in public life. People still call the clock in their millions each year, even though we have plenty more ways of finding out the time these days. The voice on the end of the line is part of our lives, and I wanted to understand why we feel so close to it in ways we never did to other recorded information lines. To do this, I needed to understand exactly why Ethel Cain won and Mary Dixon, and the other women, lost. And what I found out surprised me.

Of course, it should all have been down to the voices, and John Mase-field, the poet laureate, was quite clear that the tiny trace of a northern accent in Dixon's speech was enough to put him off. But I had a couple of doubts niggling away at me. I kept encountering news reports in the archives from 1935 that described the competition winner, Ethel Cain. There was almost blanket coverage of the contest across the national and local press, and there cannot have been many Britons who did not read about Cain's success. And I started to spot an intriguing common thread.

Here are just three reports I have picked at random. The *Croydon Advertiser* said that "She is twenty-six years of age, slim, fair and attractive." The *Evening News* described Cain as "26, dark blonde, slim, self-assured." The *Croydon Times* claimed she was "twenty-six years old, slim and fair-haired."[6] In fact, I could have picked any number of news reports from June 1935, because almost all described Cain in this way. It was almost as if her looks were more important than her golden voice. I began to wonder. Could there have been another reason, besides their voices, why Mary Dixon, aged forty-two and with stern features, lost the competition, and twenty-six-year-old Ethel Cain, with a wide, beaming smile, won?

———

THE IDEA OF staging an *X Factor*–style competition to find the voice for the speaking clock came from Stephen Tallents, a public-relations pioneer who virtually founded the modern documentary film movement and introduced PR (we might now say "spin") to the heart of government. Tallents worked under Lord Reith at the BBC and his work greatly influenced what became the Central Office of Information. He was also a powerful *éminence grise* behind the 1951 Festival of Britain, a government propaganda triumph in the aftermath of the devastation brought by the Second World War.

Before all that, Tallents had worked at the GPO, where his job was to convince the UK's population to subscribe to the telephone. This seems an easy sell, as many of us cannot imagine a world without phones today. But in the early 1930s it was only businesses which were taking up the telephone service in big numbers. Domestic use was sluggish, and Tallents knew he needed to do something more than marketing. He needed to change the public's entire relationship with the telephone. He needed to bring the telephone into the family home morally and emotionally as well as physically.

The UK was not the first country to develop a telephone speaking clock. That accolade went to France. In 1933, the Paris Observatory installed a device that gave observatory time to up to 12,000 callers each day, voiced by a Parisian radio celebrity, Marcel Laporte. In 1934, the Dutch schoolteacher Cornelia Hoogendam voiced a speaking clock at The Hague. In 1936, the famous stage and screen actor Lidia Wysocka recorded a speaking clock for Poland.

When it was first mooted by Post Office engineers who had seen the Paris clock, the idea for the UK's clock was to give it a man's voice, but then Tallents turned up and spotted a golden opportunity to get a publicity smash hit by casting a woman in the role and making her a celebrity. Tallents insisted, throughout the contest, that the voice was all that mattered. He said, "The sole purpose of this competition is to select the very best voice available, and no other consideration should be allowed to interfere with the conduct of the tests."[7] So why, just before the London competition final, did Tallents's deputy write to each of the nine finalists asking for a recent photograph of themselves? Surely, it did not matter what the women *looked* like. And why, one wonders further, did the London exchange manager include a note with one of the photos out of the four finalists in his area? It was with Ethel Cain's picture, and the note read, "Here is Miss Cain's photograph, *which I think you will like.*"[8] He did not say that about any of the other finalists.

Evidence was starting to stack up that something interesting was going on here. One newspaper commented that the judge acting as the perfect telephone subscriber should be female, "the attitude of a male subscriber to a female telephonist not being deemed as always due to telephonic considerations."[9] I also discovered a speech made by the GPO engineer who designed the clock, who carefully explained that "In view of the possibility of certain members of the public becoming so enamoured of the golden voice that they are impelled to listen to it for an indefinite period, an automatic device disconnects the circuit at the end of three minutes."[10]

Then I spotted a further clue I felt might be worth following up. In a few of the news reports about the golden-voice competition final in London, readers were told that, later in the day, Ethel Cain was taken to the nearby Prince of Wales Theatre, where she "addressed the audience."[11] Now, I could imagine what she *said*. Probably something that concluded with her exclaiming, with a great big smile, "If you want to know the time, there's no need now to ask a policeman. Just give *me* a ring sometime." That is what she said at one public appearance recorded by Pathé newsreel cameras.[12] But I wondered who she would have been saying it *to*. So I asked the archivists at the V&A Museum's theater collection to dig out the Prince of Wales theater program for that night, hoping it might give me some clues. I was not disappointed.

On June 21, 1935, the Prince of Wales Theatre, proclaiming itself proudly as "London's *Folies Bergère*," was showing *La Revue Splendide*, described as a "non-stop French revue in English."[13] In it, the barely clad showgirls performed a suite of musical numbers, including one entitled "We'll Show You a Thing or Two" and another called "Let's Start Life in a Tiny Tent." Theatergoers could hire opera glasses from attendants, and the program made a point of advising customers that "To safeguard your health—this theatre is disinfected with Jeyes' Fluid."

Non-stop revue was a craze sweeping the theaters in the 1930s, and at the Prince of Wales it went by the slogan "Come when you like—Go when you please! But No Re-Admission." One reviewer explained the sort of people that went to see this type of show: "the not-too-tired business-man who enjoys broad humour and has an appreciative eye for the female form." That was the audience that Stephen Tallents got Ethel Cain to address, the day she won the golden-voice competition. I now felt I had enough circumstantial evidence to support my hunch that Ethel Cain had been exploited by the GPO's public-relations department to act as a figure for men's sexual fantasies. Then I was shown a GPO publicity film from 1939.

The plot of *At the 3rd Stroke* revolves around a wife and husband fighting each other. I still find it hard to watch, because it depicts an abusive relationship and does so in a viciously misogynistic way. What it offered was crucial further evidence about the way the GPO publicity department was carefully creating an identity for the speaking clock in 1930s domestic life—an identity that spoke to its primary target audience.

Let me give a very brief synopsis of the film's plot. The man comes home blind drunk at four o'clock in the morning. The woman is furious at the state he is in. The man thinks the woman is a nag and provokes her aggressively into an argument. She slaps him in the face and says, "Do you know what time it is?" He slurs, "I'll call someone who'll tell me" and dials the voice of Ethel Cain's speaking clock, before settling back with a smile on his face to listen. Then he says, "A beautiful mellifluous golden voice. She's always willing. Always cheerful. Always golden. She doesn't complain when you ring her up in the middle of the night. Oh no." Then he rings the clock again, telling his wife, "I'll ask her out."

To the clock, he says, "Four-ten? It's a date. May I listen to you again?"[14] Then he falls asleep, satisfied, listening to her voice.

It hardly takes much analysis to work out what this all meant. The message from the GPO to its bill-paying subscribers, who were almost all men, was this: Do you want to escape your mundane life for a little while and experience something more . . . exotic? Well, we have just the thing. You can call Ethel Cain. You know what she sounds like and *you know what she looks like*. She is twenty-six years of age, slim, fair and attractive. And she is *always willing*. In its first year, the speaking clock, with Ethel Cain's voice, took twenty million calls. This was not just a story of the GPO sexing up a service with salacious marketing. It could have done that with any number of its other telephone services but did not. There was something specific, and highly significant, around the fact that this was a *clock*.

THIS SEEMS A good time to ask a question that may help us navigate a course through the complexity of the stories we have explored so far in this book. What *are* clocks? Perhaps the hints are adding up to something by now. Clocks are *us*, or, rather, they are proxies—stand-ins—for us and other people. We imbue clocks with identity. The telephone speaking clock made by the GPO in 1935 was not a machine, it was Ethel Cain. By dialing the right number, you could speak to *her*. You developed a relationship with *her*.

We have always felt close to the timekeepers that are by our side as we make our way through everyday life. Today, the world's biggest watchmaker, Apple, has given us a wristwatch that knows the most intimate details of our lives and bodies: our health, our fitness, our menstrual cycles. The watch has a personality called Siri. We chat constantly. "Hey Siri," we say. In time, Siri comes to know all our wants and needs, and helps meet them. Siri becomes part of the family—a parent figure.

Other watch companies offer us different, but no less intimate, experiences. The venerable firm of Charles Frodsham and Co. was once a leading chronometer maker to the British Admiralty but is now the manufacturer of carefully handcrafted mechanical wristwatches made entirely within its small workshop in Sussex. Wearing a Frodsham watch today is to borrow the identity of its team of makers, many of them young, who hold an astonishing range of craft skills handed down through generations. I often ask one of the company's directors, Richard Stenning, how his watchmakers learn the techniques and processes needed to make such fine watches from scratch. He just smiles and tells me it is a combination of an unshakable vision, determination and sheer hard work. I like to be surrounded by people like that. I see *them* when I strap a Frodsham watch to my wrist, as it gently ticks away. They are *in* the watch because they skillfully breathed life into the metal, sapphire, ceramic and glass of its components. The ticks of the watch are the very heartbeats of the craftspeople who made it.

Lewis Carroll knew about this human relationship with the time kept by clocks and watches when he wrote *Alice's Adventures in Wonderland* in 1865. Alice and the Hatter were having a tea party and the conversation got around to time. Alice was concerned that they might be wasting it, which got the Hatter irritated.

"If you knew Time as well as I do," said the Hatter, "you wouldn't talk about wasting *it*. It's *him*."

"I don't know what you mean," said Alice.

"Of course you don't!" the Hatter said, tossing his head contemptuously. "I dare say you never even spoke to Time!"

"Perhaps not," Alice cautiously replied: "but I know I have to beat time when I learn music."

"Ah! that accounts for it," said the Hatter. "He won't stand beating. Now, if you only kept on good terms with him, he'd do almost anything you liked with the clock. For instance, suppose it were nine o'clock in the morning, just time to begin lessons: you'd only have to whisper a hint to Time, and round goes the clock in a twinkling! Half-past one, time for dinner!"[15]

Watchmaker Daniela Toms adjusting a Charles Frodsham and Co.
wristwatch, 2020

It may seem nonsensical to pay too much attention to passages like this, but I think they shine a light on why we treat clocks in the ways we do: because they contain our identity. The speaking clock could take on the identity of a close friend or lover. The Edwardian electric time systems in Brno and elsewhere took on the identity of our parents, community leaders or moral guardians, admonishing us when we misbehaved and encouraging us to do better. Or perhaps they were factory managers, or mill owners, or aristocrat landowners, or colonialists, or men seeking to rule the world and to exclude women from its controlling bodies, committees, clubs, boardrooms and networks. Sometimes people resisted this control that clocks had over their lives, so they fought back. By lashing out at clocks they were really fighting the people the clocks were standing in for, and some people were prepared to kill, or to die fighting.

In Amsterdam's Stock Exchange in 1611, the tower clock was the market inspector, and the clock in the modern Tower Hamlets data center, which stamps financial trades a million times each second, has the same role. Or, rather, these clocks take on the identity of people who do not trust financial traders always to act by the rules. Those people are often called regulators, which has a certain irony, because regulators, as we have seen, are also a type of clock, namely the high-precision clocks used at astronomical observatories to time the passage of the stars which defined time for humans in the first place.

At Jaipur, the sundial at the astronomical observatory acted like a faithful, trusted and wise adviser who had traveled the world and built up impressive contacts useful to the king. The elaborate time signals constructed by British, Portuguese and French imperialists on Africa's coastline were, in reality, white overseers, standing ready to crack the whip each day and force indigenous black people back in line. Those clocks, whistles, balls and guns were violent proxies for the imperialists who hacked their way through Africa, Asia, Australasia and the Americas, seizing what and whom they wanted and letting the rest burn.

In Ajmer, Athens, Beijing, Kyoto, Melbourne and Rome, clocks raised high on columns and towers *were* the emperor. The Ajmer clock even wore the crown of Queen Victoria, the Empress of India. In Baghdad, Damascus, Diyār Bakr, Lübeck, Prague and Strasbourg, the elaborate astronomical automaton clocks were simulacra of God's universe, in which mortals formed merely the tiniest cogs, although, if all those cogs started to mesh together, they could exert great force. The clocks kept them subdued. Gog and Magog, whether in London or Michigan, stood in for all the people who have been restless for change, even if that means making enemies. In Siena, then later displayed on the walls of the finest art galleries and in the rudely carved headstones of windswept graveyards, depictions of sand clocks took on the identity of life, and then death. *Our* lives; *our* deaths.

SOMETIMES, CLOCKS STAND in for nations, which is why nationalists like talking about clocks so much. For the American pastor and orator

Thomas Starr King, whose speeches during the 1860s American Civil War stirred the nation to support Abraham Lincoln's Union, clocks played a defining role in the American national identity. At a Fourth of July address at a Sunday school in San Francisco in 1860, King hit his audience with a rousing statement about the nation's character during the eighteenth-century American Revolution:

> God does mark the great seasons of the world's history by a mighty clock. In fact, every nation has a huge dial-plate, and behind it are the works, and below it is the pendulum, and every now and then its hands mark a new hour. Our revolution was such a period . . . But the old time-piece kept ticking, ticking, the wheels kept playing calmly, 'til, about 1775, there was a strange stir and buzz and clatter inside the Case, the people couldn't bear any more, a sixtieth minute came, and all of a sudden the clock struck.[16]

If, as King preached, the American nation was like a clock, then the next step was for clockmakers to manufacture monumental devices that told America's story in *real* clockwork.

From the 1860s to the 1890s, about two dozen enormous clocks celebrating America's history were built, touring the country and overseas. One was made in about 1890 by a Boston clockmaker. Not much is known about its origins or nineteenth-century life, except that it toured Australia and New Zealand as well as the USA as part of a traveling blackface minstrel show known as Bent and Bachelder's Anglo American Christy's Minstrels. In the twentieth century it moved to a clock collector's New Hampshire barn, where, in the early 1980s, it was discovered by Carlene Stephens, who curates the clock collection at the National Museum of American History in Washington, DC.

When I was shown the thirteen-foot-high Great Historical Clock of America as the centerpiece of an NMAH exhibition on American democracy during a visit to Washington in 2018, I found it surrounded by excited visitors poring over its lavish automaton scenes and figures. Artifacts nearby included the desk used by Thomas Jefferson to write the Declaration of Independence in 1776, a newspaper printing press owned by Benjamin Franklin, and a horse-drawn wagon used by suffragists campaigning for women's rights. With such powerful icons of American identity as its backdrop, the clock commanded rapt attention, presenting a carefully constructed narrative of American history clearly influenced by the 1860s movement to associate American progress with time, and with the American Revolution.

It is a stunning show of national pride. Its clockwork mechanism causes a model of the world's first commercially successful steamboat, made by the American engineer Robert Fulton, to travel across the water, and the great Niagara Falls to cascade. Monumental symbols of America, including the Statue of Liberty and the Soldiers' National Monument, flank the clock's central displays of time and astronomical indications. Iconic figures from the American national identity, including Christopher Columbus, William Penn and Pocahontas, are shown, animated, in the clock. Near the top of the great structure is an automaton parade of every American president up to Benjamin Harrison, led by George Washington. And displayed proudly on the clock's plinth is the American flag, with the nation's motto, E Pluribus Unum, picked out in gold: out of many, one. I tried to put myself into the mind of a patriotic American who might have seen the clock on its 1890s tour, with the memory of the 1860s Civil War still painfully recent. Its message must surely have been that the United States was the greatest nation in the world.

America had not, before the war, tended much to glorify its past, preferring instead to focus on the present and future. The past had been associated with the Old World and its oppression, decadence, decay and ruin. The New World had grown up to be forever youthful. But the bloodshed of war, combined with global shifts in politics and society, had shaken the confidence of the American people, who began to yearn for their past, even if it was idealized. The Great Historical Clock of

America and its siblings, with their attractive automata of colonial and revolutionary episodes, helped Americans rebuild their sense of national identity. And they pulled off a clever trick. They glorified the past while shifting the narrative from the Old to the New World.

The clock on show in the Washington museum, like its counterparts, drew direct inspiration from the fourteenth-century Strasbourg Cathedral clock. But the American clockmakers found ways to pass judgment on Europe as they helped forge a new identity. Strasbourg's original clock was surmounted by a metal cockerel, which faithfully flapped its wings, yet sitting on top of the Great Historical Clock of America was an eagle, the symbol of the United States. The maker of a similar clock in Detroit boasted that "those who have made the pilgrimage to Strasburgh Cathedral and stood breathless in the shadow of the once king of clocks, are united in pronouncing the American Astronomical Clock superior beyond contrast to the great clock of Europe."[17] And a clock made by an Ohio jeweler in 1893 contained within itself a small-scale working model of the Strasbourg clock. In the words of Carlene Stephens and the historian Michael O'Malley, this maker "implied that Europe had been understood, captured, and surpassed—his clock displayed the Strasbourg replica like a trophy."[18]

With its parade of elaborate and monumental clocks, the New World felt like it was truly striking back, not just at its colonial past, but at its own more recent troubles. Perhaps, given its location in a museum at the heart of Washington, DC, the Great Historical Clock of America is still performing this role today.

———•———

CLOCKS, LIKE FLAGS and anthems, are deployed by leaders and governments to tell the world who their friends and enemies are. They are used to bring some people together and exclude others. In 2015, North Korea moved its standard time zone—the time shown on every clock and watch in the nation—to thirty minutes behind the time of its neighbors, South

Korea and Japan, its news agency claiming that "The wicked Japanese imperialists committed such unpardonable crimes as depriving Korea of even its standard time."[19] It was a clear symbolic statement of political allegiance and defiance. Three years later, with the relationship between North and South Korea appearing to thaw during summit talks, the North Korean leader, Kim Jong-un, agreed to return his country's clocks to the time kept in the South. He said it was "a painful wrench to see two clocks indicating Pyongyang and Seoul times hanging on a wall of the summit venue."[20] Once again, all the clocks and watches had to be shifted.

North Korea was by no means alone in signaling its political priorities and allegiances using clocks. In 2007, Venezuela's president, Hugo Chávez, had moved his country's standard time back by thirty minutes, putting the country into its own unique time zone. He claimed this was to give schoolchildren more daylight in the mornings, but two analysts, Douglas Schoen and Michael Rowan, have suggested the move was part of Chávez's program for a new national identity:

> He declared his 2000 election to be a revolution, giving him a mandate to change everything in Venezuela—the constitution, the time zone, the currency, the national shield, the national holiday, the military salute (which became "Fatherland, socialism or death"), even the country's very name—for all time.[21]

Nine years later, his successor, Nicolas Maduro, changed the clocks back, claiming the move would save electricity as the Venezuelan economy tanked.

Kim Jong-un and Hugo Chávez were just the latest national leaders to use clocks to establish authority and instill a shared sense of identity in their troubled nations. In 1949, following the country's revolution and the establishment of Beijing as the new capital, the People's Republic of China's Communist Party chairman, Mao Zedong, standardized the whole country's time to Beijing Time. It was one of his very first acts. Yet this was despite the country covering five one-hour time zones and Beijing being situated toward the eastern end of the country. "China must have a Chinese time standard," Zedong claimed.[22] By doing so, he was playing the centralizing game of identity politics that deployed clocks as proxies for the state.

The country's time became part of its national identity and keeping Beijing Time was a patriotic act. In a short story by Yun Yu entitled "Beijing Time," written in 1974 during Mao's Cultural Revolution, Maoist Red Guards used the rallying cry, "Beijing Time is the time by which Chairman Mao directs the victorious progress of the whole country!"[23] In 2001, a Chinese-state-sponsored children's magazine published a song also entitled "Beijing Time," which instructed readers to see "Beijing Time wafting on the morning breeze to the smiling face of the nation."[24] A unified time symbolized a unified nation, whatever the reality on the ground.

———•———

BETWEEN 2010 AND 2012, the UK Parliament debated a proposal to change the country's time zone by one hour from Greenwich Mean Time to Central European Time, the time kept in France, Germany, Poland, Albania and over two dozen other European countries. However, this is not how it was described. Instead, the proposed change was called "Single/Double Summer Time," and the time in summer, which would be two hours ahead of GMT, was often described as "Double British Summer Time." One might be forgiven for thinking that the clock change would make the UK somehow more British, rather than closer to its European neighbors. The proposal was dropped, but it was a foretaste for events that were to come later, when clocks returned to the front line of British politics.

———•———

ON JANUARY 31, 2020, the UK left the European Union. It was a time of great political strife, with bitter arguments about Britain's identity raging in the press, in Parliament and on the streets. Historians might pick from any number of case studies to describe the identity crisis that swept the nation, but one episode seemed to encapsulate the deeply polarized politics of Brexit more than most, and it involved (needless to say) a clock.

The striking of Big Ben, the Great Clock of Westminster, had been silenced in 2017 so that the clock and its tower could be refurbished. As the days and hours before the UK's withdrawal from Europe ticked down, a group of Brexit-supporting MPs demanded that Big Ben be reinstated so that it could toll defiantly to mark the momentous event. The Conservative MP Mark Francois led the campaign, assisted by such political heavyweights as Iain Duncan Smith, John Redwood, Nigel Farage, Matthew Hancock, Jacob Rees-Mogg and even the prime minister himself, Boris Johnson.

Speaking imperiously in the House of Commons on January 9, 2020, Francois exclaimed:

> we will leave the European Union at 11 pm GMT on 31 January. As we leave at a precise specified time, those who wish to celebrate will need to look to a clock to mark the moment. It seems inconceivable to me and many colleagues that that clock should not be the most iconic timepiece in the world, Big Ben.

He concluded that "Big Ben should bong for Brexit."[25] Arguments raged in the UK press, backward and forward, for days if not weeks. A moderate position seemed impossible: you were either for Big Ben or you were against it. People who did not care either way, and thought that the whole concept was ridiculous, were shouted down.

In the end, rows over funding and costs meant the plan to ring in the moment of Brexit ground to a halt. Big Ben remained silent on the evening of January 31, as work on its refurbishment continued. Brexiteers, furious at what they described as "a Remainer plot," found other ways to mark the moment.[26] Boris Johnson banged a small gong inside 10 Downing Street. The journalist Ian Dunt reflected that "They spent the whole week talking about bongs on Big Ben. The sheer scale of the lunacy is difficult to fully comprehend. What you are seeing, more or less in real time, is a nation turn into the clown car model of itself."[27]

It should have come as no surprise. Big Ben symbolizes Britain itself. Big Ben *is* the nation, or at least that is what some people in the UK believe. For nationalists, symbols such as national clocks and time zones are as important—sacred, even—as national flags and national anthems. An attack on the nation's clock (as it was characterized by those people who supported the Big Ben bong plan) was an attack on the British identity—an attack on the British people themselves, on what it means to be British. It was an attack, somehow, on *Britishness*.

War

Miniature Atomic Clocks, Munich, 1972

The original plan was to fly the tiny clocks to the Moon and back on the Apollo 17 lunar mission. Scientists and astronauts alike were eager to test the effects of Albert Einstein's theories of relativity, which predicted that clocks run fast or slow when traveling at different speeds and in different gravitational fields. "In fact, there is a feeling," exclaimed one clock manufacturer in 1972, "the relativity project could be among the most exciting of scientific adventures yet undertaken in space."[1] But the effects would be so slight that only the most precise and accurate clocks would show them up.

The problem facing scientists in the Apollo era was that the most accurate clocks available to them—atomic clocks, which used the properties of atoms to keep time—were bulky, fragile and power-hungry. Conversely, the only clocks small and light enough to fit into spacecraft—quartz clocks—did not offer the accuracy needed during a space voyage. But Ernst Jechart and Gerhard Hübner, working in the basement of Jechart's house in the German city of Munich, had a plan to solve the problem. Founding their own company called Efratom in 1971, the

Efratom miniature atomic clock, backup for the two clocks installed on the
NTS-1 satellite, made c. 1972

two engineers began building miniature atomic clocks, each of which
occupied a mere four-inch cube, weighed less than three pounds and
needed hardly any electrical power to operate. Introduced in 1972, they
were the smallest atomic clocks that had ever been made, and they held
out the promise of being able to test Einstein's theories in the laboratory
of space. The problem of super-accurate clocks fit for space appeared to
have been solved.

Everything had been in place to accommodate the clocks in the Apollo
17 spacecraft, but just eight weeks before it took off from the Florida
launchpad, in December 1972, NASA managers decided to pull the plug
on the clock experiment. It was a crushing disappointment that felt like
the end of a dream. But a few weeks later Jechart and Hübner received
some unexpected visitors.

Robert Kern and Arthur McCoubrey, who ran their own atomic clockmaking business in Massachusetts, were working with the US Naval
Research Laboratory on a military satellite, named NTS-1, that was due
to launch in 1974. What they needed were two miniature atomic clocks
that could be fitted into the satellite's compact payload bay, but they had
no time to develop the machines themselves. Then they heard about the

German clocks by Efratom, left behind on NASA's Apollo 17 launchpad. It was the breakthrough they were looking for. Kern and McCoubrey made a quick trip to Europe to buy the clocks, brought them back to the USA, and hurriedly modified them for incorporation into the satellite. In July 1974, NTS-1 went into orbit around the Earth, with the Efratom clocks on board. It was an early trial for America's Navstar Global Positioning System—GPS—and the two miniature clocks made in a suburb of Munich became the first atomic clocks in space.

———•———

CLOCKS HAD EXISTED at the very heart of navigation since the 1750s, when John Harrison's marine timekeepers first proved that timekeeping technology, combined with astronomical observations, offered a powerful way for military and merchant navies to find their location during voyages across the seemingly featureless expanses of the oceans. Two centuries later, the same problem had come around again, but this time with a new twist. In the 1950s, the American navy was building a ballistic-missile system as part of its military defenses against the Soviet Union. Once developed, the missiles could be aimed very precisely at their target, except for one big challenge. They were launched from a fleet of Polaris submarines, which meant that the accuracy of missile strikes depended on knowing exactly where each launch submarine was. But the guidance systems of the submerged submarines were not precise enough and, of course, they could not see the sky to read the Sun, Moon and stars. What they needed was something that could give the vessels a position fix quickly when they surfaced.

The answer came after the pioneering launch of the world's first artificial satellite, Sputnik, by the Soviet Union in 1957. American scientists tracking the satellite using radio waves realized that the same system might operate in reverse: if they knew exactly where the satellite was in space, then they could find their own position on Earth. By the 1960s, two US Navy satellite navigation systems were successfully in operation, Transit and Timation, as well as an Air Force project known as 621B.

Slowly but surely, the Pentagon started adding up the benefits of these experimental projects, and saw how they might pave the way, technologically and politically, for a bigger and more ambitious scheme. "By 1972," explained Bradford Parkinson, who led the GPS project, "a few Pentagon authorities had recognized that a new satellite-based navigation system would be a valuable asset with multiple military applications. Literally hundreds of positioning and navigation systems in use by the DoD [US Department of Defense] were expensive to maintain and upgrade."[2] During a three-day workshop held at the Pentagon over Labor Day weekend in September 1973, what was to become GPS was laid out in detail, bringing together the best parts of the Navy and Air Force approaches and defining the principles of the new global system. That December, the Pentagon gave its approval for GPS to proceed.

Crucial to the success of the scheme was the deployment of tiny, space-hardened atomic clocks on to the fleet of satellites. Each GPS satellite carries several such clocks on board, and together they beam a highly accurate time signal down to Earth at the speed of light. GPS receivers look for the time signals broadcast by four different satellites. Each is in a different part of the sky and the distance from each satellite to the GPS receiver therefore differs. This means that there is a tiny time difference between each of the time signals reaching the receiver, and those minuscule time differences are all the receiver needs to work out where it is on Earth using a mathematical process known as trilateration. The clock in the GPS receiver itself does not need to be very accurate, because it gets all its timing information from the satellite signals, and that means receivers can be made cheaply and easily. But what is crucial is that the atomic clocks carried by the satellites are *incredibly* accurate.

The first satellite navigation trials, in the 1960s and early 1970s, used quartz clocks, but they were not accurate enough to give the precision that military users needed. Once Ernst Jechart and Gerhard Hübner's miniature atomic clocks became available in 1972, it seemed that the problem had been solved, and the 1974 NTS-1 satellite trial that carried their clocks went well. But engineers knew that the clocks destined for GPS satellites would need to be much more rugged to keep ticking without fail for years on end as the satellites orbited Earth. It was like the eighteenth-century longitude problem all over again. Then, John Harrison had to find ways to miniaturize existing timekeeping technology while also increasing its accuracy, precision and long-term reliability. At the same time, he needed to harden his new timekeepers against the hostile conditions of a long sea voyage such as violent movement and temperature fluctuations. Clocks for space had to deal with the added challenges of radiation, vacuum and gravitational forces ranging from high, during take-off, to zero, when orbiting.

In 1973, before the NTS-1 trial had even launched, Jechart and Hübner's Efratom company opened a factory in California, close to Rockwell, the company which, in 1974, was contracted to build the fleet of GPS satellites. Together, the two firms set to work hardening Efratom's pioneering miniature atomic clocks for the GPS program. And one by one, from 1978, GPS satellites carrying the newly rugged atomic clocks began taking to the skies, launched by old American intercontinental ballistic missiles that had been modified for the purpose. Over the next seventeen years, more than thirty GPS satellites went into Earth orbit until, by 1995, the first phase of the system was complete, and GPS was declared fully operational. These clocks were destined to change the face of war, and of everyday life, as clocks have done for centuries, but this time things would change in ways we still cannot fully comprehend.

———•———

CLOCKS ENABLE WAR, and war shapes the way we use clocks in return. To pluck just one example from history, consider how today's military rockets and missiles owe a debt to William Congreve, a pioneer in rocketry who, in 1808, designed a new type of clock to time rocket flights. Congreve's clock involved a tiny metal ball zigzagging along a track cut

Thwaites and Reed rolling-ball clock, made c. 1972

into a tilted metal plate until it reached the end of the track, triggering a spring to flip the plate the other way and repeating the process. Congreve held out great hopes for the accuracy of his invention, but it was flawed in engineering terms and turned out to be a poor timekeeper. But it was soon followed by other horological innovations built to serve military ballistics and the precise delivery of explosives. In the 1970s the British clockmakers Thwaites and Reed made replicas of Congreve's clock that can now be found in many museum collections as an engaging horological curiosity from the past. By viewing these curious clocks we are seeing the history of brutal warfare.

Or, we might think about the work of companies such as Thomas Mercer, who made the marine chronometers that helped navies navigate safely as they waged wars and built empires, but whose business started to dry up in the Cold War as radio location aids took over. So Mercer moved with the times. The nuclear warheads carried on the UK's Polaris missiles needed detonation timers so that they would explode just before the missile hit the target, maximizing destruction and loss of life. But electronic timers would be destroyed by the massive electromagnetic pulses caused by the detonation of other missiles in the bombing raid. Battle-hardened mechanical timers developed in 1982 were what Mercer contributed to the nuclear age. These, too, can now be seen in museum displays, though their modest form means they are often overlooked. We should look more closely.

In 1947, when Martyl Langsdorf designed "The Clock of Doom" for publication in the *Bulletin of the Atomic Scientists*, her symbol of nuclear anxiety—a clock set at seven minutes to midnight—represented a world that felt it was fast running out of time. Since then, the hand has frequently been moved to reflect changing concerns over impending Western apocalypse. On January 23, 2020, the *Bulletin*'s advisory board moved the hand of the clock to 100 seconds before midnight—its closest position yet. Clocks teach us about nuclear Armageddon because they remind us what happens when time runs out.

We could also look beyond individual clocks and consider how whole classes of timekeepers were shaped by the requirements of war, and in turn shaped us. The wristwatches that many of us wear every day owe a strong debt to war. During the South African War of 1899–1902 and the First World War of 1914–18, soldiers began to strap pocket watches to their wrists so they could time the waves of attacks while keeping their hands free to wield weapons. Wristwatches existed before then, as women's jewelry or used by women in pursuits such as cycling and horse riding, but war turned wristwatches into unisex products, doubling the market and quickly causing pocket watch manufacture to fall into terminal decline. Today's multibillion-dollar wristwatch industry is built on the back of two brutal wars.

We might even zoom right out and see ways in which the temporal patterns by which many of us live our lives have been directly affected by the demands of war. War is indelibly imprinted on the horological practice that a quarter of the world's population carries out twice each year. Earlier, we explored Daylight Saving Time, the practice of advancing clocks and watches by an hour in summer in order to shift the hours

Doomsday Clock, photographed on January 23, 2020, after having been adjusted to 100 seconds to midnight

we are awake a little earlier in the day. It was war that turned this idea from a civilian oddity into a military necessity, in both the First World War and the Second World War, with munitions factories in full production and fuel for lighting and power in short supply. Here, in the most widespread way, we see clocks enabling efficient wars and then reshaping the patterns of peacetime.

These and countless other examples simply remind us that technological development is often catalyzed, accelerated or shaped by war, and clocks sit at the heart of it all. Clocks have always been as much a class of military weapon as are bullets and bombs. But GPS has weaponized clocks like no other military project in history, so I feel it is worth getting to know the satellite clockmakers a little.

The experimental Efratom clocks used in the NTS-1 trial, made largely in Munich, continued to be developed for the first block of GPS satellites by the Rockwell and Efratom companies in Irvine, California.

Clocks in later GPS satellites were made by a variety of different US firms. But GPS is not the only fleet of navigation satellites orbiting our planet. Russia's global system, called GLONASS, uses clocks made at a facility by the Neva River in St. Petersburg. Europe has a network called Galileo, whose satellites carry clocks made by firms based in Neuchâtel and Rome. China's system, known as BeiDou, originally used clocks bought from a Swiss maker, but later utilized its own clocks made by the China Aerospace Science and Industry Corporation.

Two smaller, regional, navigation systems exist, too. India's IRNSS uses satellite clocks made in Germany, though, like China, it later used home-grown clocks, in its case developed by the Indian Space Research Organization at Ahmedabad. Japan's regional network, known as QZSS, uses clocks made in Massachusetts.

It is a truly global network of clockmaking for a global constellation of navigation satellites. Together, over 100 of these satellites are orbiting our planet right now, beaming precise time signals down to Earth from something like 300 miniature atomic clocks made in these factories across the world. Not only do they give satnav receivers their precise location, but they are also used to set countless other clocks right. They are now the time signal for the world. The time kept by these 300 or so clocks reaches more human beings than any other clocks that have ever been made. We should therefore know and understand them—not so much how they work but what they mean. Because these clocks, conceived in the Cold War of the twentieth century, are part of a global reimagining of war in the twenty-first century: they are changing what war means forever.

———

ARMED CONFLICT IN the twenty-first century is not like the wars of the twentieth century, whether hot or cold. Then, boundaries seemed clear. One nation, or a group of nations in an alliance, fought another. The wars were existential—the continued existence of the warring states depended on winning. Wars, and truces, were *declared*. They had start and end dates. They had *names*. Today, postmodern warfare is not so clear-cut. The enemy may not be a definable nation or state. It might not even really be definable at all. Who is the enemy in a war on terror? Who is fighting whom? When will we know it has stopped—if it ever does? This is not just a matter of dissolved national boundaries. In today's wars, the people doing the fighting may not even be the official armed forces employed by the states or groups involved. War is often now framed in the murky language of human "surrogates," which the defense scholars Andreas Krieg and Jean-Marc Rickli describe as "terrorist organizations, insurgency groups, transnational movements, mercenaries or private military and security companies."[3] Armies build alliances with local "partners," often on shifting sands, to contract out much of the actual fighting.

Nor is the battlefield itself easy to define. Where does war take place? It is no longer confined to the territories of the warring states, even if these can be clearly defined. The electronic technologies of mobility, navigation and telecommunications—guided missiles, drones, robots, satellites, mobile phones, the internet, social media—are taking the fight to new places. Places like high-altitude airspace, where drones can live; space itself, where satellites live; and cyberspace. Cyberspace is especially fertile territory for surrogate war because it is almost impossible to unravel who is really who.

What does this mean in practice? America's "war on terror," following the 9/11 attacks, is often considered a war waged in Afghanistan and Iraq, but US military activity in the decade since 2001 was also directed into places including Iran, Libya, Pakistan, Somalia, Yemen and Mexico. Moreover, fighting in those battlegrounds was accompanied by singular urban attacks taking the war into the heart of cities around the world as well as the more nebulous territory of cyberspace. It was, and remains, an "everywhere war," in the words of the geographer Derek Gregory.[4]

More recently, this war has also included a fight against the Islamic State, or ISIS. In 2014, America began high-technology airstrikes in northern Syria. The following year it began supporting Syrian opposition militia groups with training, equipment and advice. No war was declared, and only a couple of thousand US soldiers entered Syria. The local partners took the burden of fighting. US exposure in human and financial terms was limited. But this was not just a fight with missiles and bombs. ISIS became ruthlessly adept at shaping narratives using the grim spectacle of video, often live, shared around the world on social media.

One ambition behind this shifting of the practice of war has been to take as many soldiers as possible out of the battlefield, in the hope of waging apparently bloodless wars. Another has been to reduce the political risk to national governments in waging controversial wars. Of course, people still die, and one consequence has been that conflict is now all around us, physically and digitally. All this change has gone hand in hand with globalization, in the increasing interconnectedness of global society and politics. The very idea of a "nation" is, in many ways, starting to break down, replaced by the idea of the "transnational."

IN 1996, ONE year after GPS was declared fully operational, the US National Security Council put out a press notice in which it claimed that GPS "is now being integrated into virtually every facet of our military operations."[5] But it had proved its worth to the American military long before the constellation of satellites was completed.

The US Navy used GPS to find minefields in the shipping lanes of the Persian Gulf in 1987, and the Air Force made use of the satellite clocks in 1989 during *Operation Just Cause*, when America invaded Panama. In *Operation Desert Storm*, the Gulf War of 1990 to 1991, GPS played a major—possibly war-winning—role. In *Operation Restore Hope* of 1993, GPS was used to target food and supply airdrops in Somalia, and the following year US troops used GPS as they invaded Haiti in *Operation Uphold Democracy*. In the first few years of the system's availability, a whole range of missiles and bombs became GPS-guided, and attention turned to smaller and smaller rockets and shells that could be fired from the decks of ships, all guided to their targets by GPS signals. The dropping of equipment and soldiers into war zones began to benefit from GPS navigation, and GPS enabled the development of a new family of weaponry: autonomous and remotely controlled robots, including drone aircraft, drone underwater minesweepers and robotic rescue vehicles to recover wounded soldiers. Friendly-fire incidents, which had killed or injured some 250,000 American soldiers in the twentieth century, were dramatically reduced with the assistance of the satellite clocks orbiting overhead.

But GPS had been a hard project to sell; America's military commanders originally struggled to see how the system would help them. One GPS project engineer involved with developing hardened atomic clocks recalled, "There were lots of arguments over whether the system would even work and we still had the old bomber pilots in the government who were just interested in dropping bombs. I had one retired four-star general asking: 'who needs it? we'll just drop a bigger bomb.'"[6] The general, who went on to oversee the Gulf War air campaign, had initially been scornful about the teams working on space-based systems: "They're paid to dream. We're paid to kill."[7] But once the system started producing results, it did not take long for even the most skeptical military general to realize that there would soon no longer be bombs, or killing, *without* the GPS clocks. Warfare was moving from indiscriminate mass killing with big bombs to targeted surgical strikes with GPS-guided weapons.

In 2004, nine years after the system was declared fully operational, the project leader, Bradford Parkinson, reflected:

> If you look at Kosovo, Iraq I, and Iraq II, I think you'll see that virtually every weapons system relied on GPS. I call it the "humanitarian bombing system." Because with a GPS-guided bomb you can hit what you're trying to hit and not hit what you aren't trying to hit.

You can work closely with your troops on the ground and not fear you're going to hit them, or a mosque, or a hospital.[8]

He later recalled:

I had great sensitivity to the fact that everything we were doing really related to the warrior. We were trying to put together a system that would enhance and revolutionize warfare. The model that we had—"Drop five bombs in the same hole"—meant, don't forget the end product of what we were trying to do here.[9]

The best GPS receivers today can calculate position to within ten millimeters—the length of your little fingernail.

But this is not merely a story of GPS-guided missiles. GPS is everywhere and that means war is everywhere. Globalization has created a web of connections between states, institutions, private corporations and individuals. Any state that fights another state risks harming itself and its interests. Warfare has followed globalization in becoming a web of connections and acts. GPS is global because there is no such concept any more as a singular American military force. Warring states are entangled, as are national and commercial interests.

When Ronald Reagan declared publicly that GPS would be made available to civilian aircraft, following the 1983 shooting down by the Soviet military of Korean Air Lines Flight 007, he was making political capital out of a decision that had been made in the early 1970s as GPS was being designed. It was always designed for use beyond the US military and it has now become a service for the globe. That is the problem.

———•———

AS OF 2020, seventy-four GPS satellites have been launched, each carrying three or four atomic clocks on board. Those GPS clocks, and those of the rival systems of Russia, China and Europe, are ticking for all of us. Now, everything from transport networks, power lines, telecommunications and surveying, to agriculture, banking, meteorology and the emergency services rely on the time, position and navigation information that satellite clocks provide. Every aircraft in the skies is navigating, right now, using clocks in space. Every ship on the oceans is making its way using satnav. That Amazon delivery you're waiting for? GPS tracks it all the way. The lights are on because of the space clocks. The computer on which I am typing this, in my flat in Greenwich a few hundred yards from the prime meridian of the world, is powered because GPS synchronizes the national electricity grid, and it connects to the web because GPS signals keep everything in step. World banking is coordinated by satellite clocks, as is TV and radio broadcasting.

Right now, experts say that 80 percent of the world's adult population has access to a smartphone using GPS or one of the other global systems. By 2022, it is estimated, there will be over seven billion GPS receivers in use around the world. That is the same number as there are humans on Earth. Almost all technological systems that rely on time have their own clocks corrected by the time signals beaming down from space. In the UK, the BBC's six-pips time signal, first set up in 1924, has been set using GPS time for the last fifteen years. The telephone speaking clock works out its time from GPS signals. Most cell-phone clocks, most laptop clocks and the clocks on most railway departure boards get their time ultimately from the clocks orbiting Earth on satellites. The hedge funders and high-frequency traders making money on the financial markets are relying on GPS, even those who pay for the higher performance that the National Physical Laboratory provides down its dedicated fiber link. Somewhere in their network GPS will be keeping time. Even NPL itself uses navigation satellites to compare its clocks with those in other countries.

GPS or other satnav receivers are embedded in virtually every mobile phone, aircraft, ship, automobile, telecommunications mast, power sub-station, TV station, data center and water pumping station. If we want faster data on our computer networks and mobile phones, we need more precise time synchronization. All digital systems, and all infrastructure controlled by computers, rely on clocks to coordinate all the data that flows over their networks. And, right now, the clocks they all use to set themselves to time are satellite clocks. You name it, there will be a sat-nav receiver locked on to the signals beaming down from those miniature atomic clocks in space. These are the networks and systems that keep us alive, with food on our plates and roofs over our heads. Without them, it is not hyperbole to say that the modern world would grind to a halt.

With all this talk of time signals beamed down from space, you might be thinking about the times your TV signal gets a bit noisy during stormy weather, or when your Wi-Fi signal at home sometimes drops out when you move too far away from the router or a couple too many walls get between it and your device. Maybe you drive for a living and know that the van radio drops out when you go into a tunnel. The point is that wireless signals are vulnerable and can easily be blocked. What would happen if the low-powered satellite time signals, coming all the way from space, dropped out or got blocked? Well, you would be right to be con-cerned, because that is what makes national governments, commercial leaders and military commanders lose sleep, too.

Satellite navigation signals are vulnerable in four main ways. The first is that they can be affected by errors in the complex system of satellite clocks. In 2016, an error of just fourteen millionths of a second in the GPS time signals, caused by a mistake made during routine maintenance, led to four days of disruption at a major UK telecommunications operator, as well as affecting critical infrastructure in Spain and America. Fourteen millionths of a second—fourteen microseconds—is a lifetime.

The second vulnerability is loss of the signal owing to natural forces. If the Sun throws out too many solar flares or you're in a city with tall buildings, GPS can be lost. It is alarming how often this happens.

But it is the third and fourth vulnerabilities of GPS and its rival sys-tems that bring civilian use of the system firmly back into the scope of war. The third is deliberate jamming, using simple electronic equipment readily available to criminals or enemy forces. In 2009, tests were carried out on a British ship using a tiny GPS jammer that had less than one-thousandth of the power of a cell phone, to see what would happen to the ship's systems. A government report later claimed that the jamming device:

> caused the electronic chart displays to show false positions. As a result, the autopilot steered the ship quietly off course. The automatic identification system reported those incorrect positions to other ships manoeuvring nearby and to the vessel traffic service ashore. The jammer also caused the satellite communications system to fail. The ship lost its distress safety system, there to raise alarms and guide rescuers. The helicopter deck stabilisation failed. Even the ship's clocks went wrong. And the usually reliable fallbacks, radar and gyrocompass, both gave warnings, as they too use GPS inputs.[10]

In 2013, at least 250 commercial airline flights near South Korea's Incheon International Airport had to switch to backup navigation systems because the GPS service they relied on was jammed by signals believed to have come from nearby North Korea. The signals were so powerful that they disrupted the cell-phone network in Seoul, thirty or so miles away, which relied on GPS clocks to work. The previous year, a New Jersey truck driver inadvertently jammed the GPS signals at Newark International Airport with a device he had installed in his Ford pickup to prevent his employers from tracking his location. A jamming detector installed on the roof of a City of London building that houses financial trading and telecommunications equipment, which requires highly precise time signals, picked up an average of five jamming incidents every day for four years, mostly caused by vehicles nearby fitted with jammers.

But the most insidious problem of all is a practice known as "spoofing." When your enemy spoofs a GPS signal, you do not know it has been affected, but your receiver thinks it is somewhere it is not. Imagine if that receiver is on a warship, or guiding a missile, or dropping your soldiers into a war zone. In the 1997 James Bond film, *Tomorrow Never Dies*, the evil media baron Elliot Carver spoofs a GPS signal that causes the British warship HMS *Devonshire* to stray into Chinese waters, almost leading to a Third World War. "Are we absolutely sure of our position?" demands the ship's captain. "Yes, sir. An exact satellite fix," confirms the navigating officer. Yet the ship was miles off.

It was a remarkably prescient fictional plot, coming fully two decades before the first real major documented GPS spoofing attack. In 2017, twenty merchant ships sailing near the Russian port of Novorossiysk, on the Black Sea, reported their GPS receivers were placing them inland, at the nearby Gelendzhik Airport. The previous year, cell-phone apps in use near the Kremlin building, in the Russian capital of Moscow, had been showing the phones' position incorrectly as Vnukovo Airport, a little under twenty miles away. Commentators described the spoofing attacks as a form of electronic warfare.

The risks involved in all this are huge. Experts working with the latest electric supply networks, known as smart grids, rely on measuring equipment that is synchronized to within billionths of a second. In one reported case, a test device locked on to a single satellite signal, rather than the several signals it needed to get an accurate fix, and the system thought there had been a major electrical fault, so it shut down two 500 kV power lines. That time it was an accident, but the risk of malicious activity is great.

Our modern world is so reliant on signals from satellite clocks that the scope for attack by jamming or spoofing them is vast. One US politician recently claimed that "GPS is the single point of failure for the entire modern economy." Another expert was more direct: if there was a widespread outage of GPS, "People will die."[11]

Peace

Plutonium Timekeeper, Osaka, 6970

I t is the seventieth century, in a city once known as Osaka, in a land that the history books say was called Japan and, after everything that has happened since it was buried deep underground, somehow the clock is still here, and still ticking. After all the war and bloodshed in the world; after so many civilizations across continents and millennia have risen and fallen, so many global empires been built and then torn apart, so much knowledge gained and lost, so much of the Earth scarred by humanity's insatiable greed for resources, this clock has survived: undisturbed, unbroken, alive. Here it has sat, quietly marking time, for 5,000 years, a silent witness to the tumultuous changes that have raged across the world above. But its time is about to run out: the hand on its circular metal dial is about to reach the fiftieth marker on the scale, where each division has recorded the passing of another century.

The clock in question is a plutonium timekeeper, made by the Japanese electronics firms of Matsushita and Seiko, housed in a polished metal cylinder. A single gram of radioactive plutonium in the form of an oxide, wrapped in gold foil, has been steadily radiating helium nuclei

Plutonium timekeeper buried at Osaka, 1970

into the clock's gas chamber, which has been expanding like an accordi-on's bellows as a result. As the bellows have expanded, they have been pulling the clock's single hand around the dial, slowly but surely.

Nobody has seen the clock since it was sealed inside a three-foot-wide, 130-gallon, spherical, stainless-steel capsule, protected by layers of steel, sand, clay and reinforced concrete, and then lowered into a hole forty-six feet deep beneath a public park in Osaka. It is fifty centuries since the timekeeper was set running after the international exhibition held in the city in 1970. But that is in the mists of time; the civilization that mounted the exhibition has long since passed into history. The idea of humans living so long ago seems unimaginable, but now there is a chance to experience something of their existence; to hold things that they held; to examine artifacts they cared about. For today, in the year 6970, the moment has finally arrived for the plutonium clock and its capsule to be unearthed, brought back to the surface, and opened.

What was sealed in Osaka on December 22, 1970, and buried on

January 20, 1971, all that time ago, is a time capsule. Besides the pluto-nium clock, it contains over 2,000 further objects—art, literature, music, preserved life-forms and artifacts—all carefully sourced and assembled in the late 1960s and buried as a message of progress and harmony for humanity in the future. There are believed to be biological specimens, artworks, tape recordings, films, books and items from everyday life held within its compartments. A 1970 version of the ancient Egyptian Rosetta Stone from 196 BCE, this one etched on a steel plate in Japanese, Chinese, English, French, Russian and Spanish, carries the message "We who believe in the prosperity of mankind 5,000 years hence bequeath to posterity this capsule as a record of the 20th century."[1] Japanese school-children made paintings for the people of the future—the people of *today*. A Japanese flag completes the inventory.

The capsule is sitting deep beneath a stainless-steel semispherical mon-ument mounted on a polished Portland granite plinth occupying almost 400 square feet at the heart of Osaka Castle Park, near a former military castle initially built in the 1580s and rebuilt from time to time since then. Early ideas for where the capsule should have been placed included an ocean, Antarctica or even the Moon, which received its first human vis-itors in July 1969, a few months after the Osaka site was chosen. But the park is where it ended up, and countless visitors have stood by the monu-ment since the capsule was first buried, reflecting on the messages it con-tains and how life in 1970 might have compared with their own existence.

That was a long time ago. Now, 5,000 years after the capsule was care-fully sealed by its makers, it is almost time to see its contents once more. The exhibition organizers decreed that it must be opened now, but it will not be easy. The sealing process was complex and technically challeng-ing, so a set of special instructions detailing the method required to open the capsule were buried above it, encased in a further stainless-steel cyl-inder. So long as the instructions have survived, and are followed, the contents of the capsule can finally and safely be revealed after their long hibernation in the Osaka earth.

At last, the hand of the plutonium timekeeper reaches its final position on the dial. Here, in a peaceful park at the heart of a great Japanese city, the time has come to reveal this message of hope and peace transmitted from a long-distant past. But a crucial question remains. Today, fifty centuries on, is there a human civilization still alive on Earth to perform this momentous act?

THE TIME CAPSULE buried in the grounds of Osaka Castle after the Japan World Exposition in 1970 was created by a society still struggling to come to terms with the aftermath of a devastating world war, ended in 1945 after atomic bombs dropped on the Japanese cities of Hiroshima and Nagasaki killed some 200,000 people and ushered the world into the nuclear age.

A sense of living on a threshold between an old and a new world suffused the project, with cosponsor the Matsushita company writing:

> Every concern of the world as a whole is a concern of Japan also. The threat of nuclear war, atmospheric pollution, dwindling supplies of energy resources and the natural hazards of earthquake, storm and flood are as real to the Japanese people as they are to the people of any other race. And the Japanese people have an equal share of happiness and sadness in everyday life.[2]

Matsushita was a major manufacturer of electrical devices during the Second World War. Matsushita Masaharu, president of the company, wrote these words for inclusion in the capsule:

> Other great and colourful civilizations have preceded ours and we are aware that our knowledge, our science and our art in the 20th century owe much to the hard-won achievements of those who have gone before. As long as mankind lives in the universe, every civilization will be a priceless legacy. I believe that something of our

civilization will survive the ravages of time. But eventually much of what we have achieved will disappear without trace—as we know from our investigations into past civilizations—due to human destruction or natural causes.[3]

As we come to the end of our journey about time, what is next for clocks and civilization? After all that we have seen about how clocks have enabled the most tumultuous and disruptive change, can we imagine ways in which we could use clocks to make peace? Time capsules such as the one buried in Osaka can be considered as clocks, but clocks unlike all the others we have examined so far in this story. Where they have been concerned with the *now*, or *nowadays*, time capsules work over longer time horizons. They work over the *long now*. And that might just save civilization.

———

IN 1998, STEWART BRAND, writer, inventor and founder of the *Whole Earth Catalog*, wrote these words:

> Civilization is revving itself into a pathologically short attention span. The trend might be coming from the acceleration of technology, the short-horizon perspective of market-driven economics, the next-election perspective of democracies, or the distractions of personal multitasking. All are on the increase. Some sort of balancing corrective to the short-sightedness is needed—some mechanism or myth that encourages the long view and the taking of long-term responsibility, where "the long term" is measured at least in centuries.[4]

Brand, together with the computer designer Danny Hillis and other prominent thinkers, had become increasingly concerned that the year 2000 had progressively acted as a temporal mental barrier to the future.

Brand asked, "How do we make long-term thinking automatic and common instead of difficult and rare? How do we make the taking of long-term responsibility inevitable?"[5]

Hillis's proposal was to build "both a mechanism and a myth" in the form of a mechanical clock the size of a great monument, capable of telling time for at least 10,000 years if it was looked after and cared about by the civilizations that had custody of it. The result was a foundation, named the "Long Now Foundation," and a project to build "The Clock of the Long Now."

The name came after the musician and composer Brian Eno suggested a concept that might extend our time horizons in a useful way. In this scheme, "now" meant the present moment plus or minus a day, while "nowadays" extended that time horizon to about ten years forward and backward. But "the long now" extended our temporal envelope massively. The futurist Peter Schwartz proposed that the long now should mean the present day plus or minus 10,000 years. "10,000 years ago was the end of the Ice Age and beginning of agriculture and civilization; we should develop an equal perspective into the future," as Brand has explained.[6]

The prototype Clock of the Long Now first ticked at the San Francisco Presidio park moments before the end of New Year's Eve 1999, two years after a small team of engineers and designers, led by Alexander Rose, had begun its construction. It is driven by falling weights, wound every few days or so. The pendulum, which twists like that of an anniversary clock, ticks twice per minute, working through an oversized watch mechanism to transmit time to the mechanical computer at the heart of the clock. This computer, inspired by the machines of the nineteenth-century inventor Charles Babbage, performs a calculation and updates the dial display once each hour. The slowest part of the dial rotates once in about 26,000 years, showing the precession of the equinoxes.

As the designer of some of the world's fastest supercomputers in the 1980s, Hillis said in the 1990s that he wanted to atone for his sins in speeding up the world by designing the world's slowest computer for the Long Now clock. And this clock has changed my life.

I first met Rose, Hillis and Brand in the year 2000, when they visited the Science Museum in London, where I worked at the time. The museum was in the final stages of building its landmark millennium display, *Making the Modern World*, and the Long Now Foundation had agreed to lend the prototype clock for long-term display as the final exhibit in the gallery when it opened in July. We could think of no better artifact to conclude our historical display and invite visitors to look forward long into the future. And there, after its journey from San Francisco to London, it stood. A mechanism and a myth—and I was entrusted to wind it up. As a young and junior curator, it was an experience that I have never forgotten.

A second clock is now being built in Texas, and a site for a third clock has been selected on a mountain range in the Nevada desert. Most of the Nevada location is covered by a forest of bristlecone pine trees, the world's oldest living organisms. One tree on the mountain range has been dated at almost 5,000 years old.

Once you start making these long focus shifts of your temporal horizon—from a few minutes or hours to a few thousand years—some astonishingly powerful ideas can emerge. Brian Eno took on the job of designing the clock's bell-chiming mechanism and started experimenting with the practice of change-ringing. He explained:

> By the 1600's, many churches had numbers of tuned bells in their belfries, and these seem to have been rung in simple sequence, over and over. About that time a new development arose. Ringers became interested in trying to ring all the possible permutations of their bells, and it is this pursuit that has become known as change-ringing. Stated briefly, change-ringing is the art (or, to many practitioners, the science) of ringing a given number of bells such that all possible sequences are used without any being repeated.[7]

The number of unique permutations of a peal of bells rises factorially as the number of bells in the peal increases. With two bells, there are two different "tunes." Add a third bell, and the number of unique tunes increases to six. Add a further bell, to make four in total, and you can ring twenty-four different tunes. With five bells there are 120 unique permutations, rising to 720 with six bells. But what happens with a peal of ten bells? It is not uncommon for a church to be fitted with a peal like this—the church of St. Alfege, close to my flat in Greenwich, has had ten bells hanging in its tower since the 1730s. If you crunch the numbers, you find that these ten bells can be rung in over 3.6 million unique sequences.

But change-ringers do not need to memorize any of them; they have a set of simple rules—an algorithm—that tells them what the next sequence is, and the next, and the one after that. The tunes are constructed algorithmically, by the set of rules. So, if you were to operate the algorithm repeatedly, you could know the tune that would be played at any point in the future. Conversely, if you heard a sequence, and knew the starting sequence, and applied the algorithm in reverse until you reached the start, you would know how far through the permutations you were.

The details do not matter here. The consequences do. It turns out that the number of unique tunes that can be played on ten bells is very close to the number of days in 10,000 years, the minimum life span of the Clock of the Long Now. What Eno realized is that a peal of ten bells attached to the clock could play a unique sequence every day for 10,000 years. And, by knowing the algorithm that changed each tune to the next, he found he had a calendar which could keep track of time over 10,000 years using just ten bells. In 2003, Eno made a CD which played the bell sequences for 5,000 years hence—January 7003. I have listened to it hundreds of times over the years and it always makes me stop and reflect. I wish I could hear the real thing, but that will have to wait.

The conversation is building. On December 31, 1999, as the Long Now Foundation set running its prototype clock in San Francisco, the artist and musician Jem Finer began performing *Longplayer*, a musical composition that will play for 1,000 years before repeating. Its permanent home is a nineteenth-century lighthouse on the River Thames in east London, though it can be heard around the world, and this strange, beautiful and beguiling music enchants all those who hear it. A few steps from the lighthouse is *Alunatime*, a towering steel, iron and glass clock designed by the artist Laura Williams that shows lunar time: the wax and wane of the Moon's phase, its rising and setting, and the progress of the tides in the river that flows alongside. It is the first experimental manifestation of Aluna, a project that will build a monumental-scale lunar clock, larger than Stonehenge. The ambition of this clock is to help unite a fractured world—a world that, despite its differences, looks up together at one Moon. Looking long into the future and far out into space helps put our earthly concerns into perspective.

Like the Long Now clock, Longplayer and Aluna are machines for making good conversations. Long Now board member Kevin Kelly has asked, "If a Clock can keep going for ten millennia, shouldn't we make sure our civilization does as well?" The virologist Jonas Salk had previously inquired, "Are we being good ancestors?"[8] And this takes us back to Japan, where we started this journey of hope and peace.

———

SEVENTY MILES FROM Osaka is the Japanese city of Ise, which is home to the Ise Jingu, or Grand Shrine, a complex of 125 Shinto shrines covering an area the size of central Paris. Occupied from about 4 BCE as a site to worship the Sun goddess Amaterasu-Omikami, the most revered ancestor of the emperors of Japan, the first shrine buildings were constructed in the seventh century. But the structures you see there today are only a few years old—and that has been the case for over 1,300 years. Every twenty years since 692, barring a few periods of warfare in the fifteenth and sixteenth centuries, the wooden shrine buildings at Ise have been totally rebuilt; the most recent ceremony, in 2013, was the sixty-second rebuilding. About 100 local craftspeople are entrusted with the process, which takes eight years each time.

In his book outlining the Clock of the Long Now project, Stewart Brand said, "Ise is the world's greatest monument to continuity—an unbroken lineage of structure, records, and tradition on a humid,

earthquake-prone, volcanic island. Its ancient rites are alive and mean-
ingful."[9] Alexander Rose, Danny Hillis and other Long Now Foundation
members attended the 2013 rebuilding ceremony. Afterward, Rose wrote:

> I watched as Crown Princess Masako led a procession of Shinto
> priests carrying treasures from the old temple to the new . . . As
> part of Shinto ritual, this not only keeps the structures intact even
> when made out of relatively ephemeral materials, but it allows the
> master temple builder to train the next generation.[10]

In Osaka, for the 1970 international exposition, the Ise rebuilding ritual helped inspire the time capsule project. In fact, the capsule that is to
stay undisturbed until the year 6970 is not the only one that was buried
there. Two identical capsules, with identical contents, lie under the monument slab. Only the lower one will remain undisturbed for 5,000 years.
The upper capsule was unearthed, opened, inspected and reburied in the
year 2000, and will be re-inspected every 100 years until five millennia
have passed. Like the Ise shrine, the Osaka capsules are a case where
repeated attention is believed to be the recipe for long-term survival.
Performance means permanence.

IN *ABOUT TIME*, we have gone back more than two millennia to the
ancient world, and we have looked forward with hope to a time 5,000
years in the future. In every civilization, and throughout time, we have
encountered clocks that people have made to advance some sort of goal—
a wider goal than simply measuring time. Or, rather, we have found that
all clocks have been made to do this. Every single clock in history.

Technology is never neutral, because objects are made by people with
an agenda of some sort. That might be an ambition to secure or maintain power, to shape moral behavior, to make money or to take over the
world. But we have also found people fighting back against perceived

injustices or inequalities, hoping for a fairer world. None of this is likely to change in the future.

But some things might. It is possible that the ways some of us live, work and travel will change. We have long since got used to rush hours, for instance, which are caused by fixed working times at fixed workplaces relying on efficient transport networks, all managed using clocks. This led in turn to ideas like daylight saving, which changed our psychological idea of time and the relationship we have with clocks. But there is no reason to suppose that we must endure rush hours forever, as changes in information and communications technologies are making flexible patterns and places of work possible for some people (though by no means all). I must say I am skeptical that we will see major, lasting shifts in our patterns of work and travel on a global scale in the coming few years, but whatever patterns do emerge, it is certain that clocks will remain at the heart of making them possible.

Related to this are potentially radical changes in the ways we make and consume. The manufacture, movement and consumption of foodstuffs, commodities and goods around the world is less often talked about in a political sense than the movement of people—that became painfully evident in the way Brexit was discussed in the UK, for instance—but the Covid-19 pandemic quickly brought the subject into sharp focus, although the changes had been bubbling up for decades. Clocks have always been at the heart of markets, finance, trade, manufacture and global transportation networks, which will not change, but the way the landscape of production and consumption is configured certainly will. Climate change is the obvious long-term existential challenge here, but we will face others.

We are also becoming, as a civilization, more conscious as time goes on of the politics of technology. We now have a wealth of evidence of the political histories embedded in artifacts. When we think about decolonization, for instance, our focus is often on the structures of inequality and injustice, which can be revealed (or are constructed) by objects. Statues and museum collections rightly come in for close attention as they invite us to question past actions critically. Perhaps we should also look more closely at political objects like clocks, which hide in plain sight all around us.

If I were to pick one clock story that worries me the most as we move into an uncertain future, it would be the clocks orbiting Earth on global navigation satellites operated by the American, Chinese and Russian militaries, as well as the European network, Galileo. In themselves, they are just flying clocks. But they enable other technologies, other networks, the exploitation of other data, and their rapidly increasing reach into all aspects of our lives—the lives of everybody on Earth, of whatever station—should give us all pause for thought. Technologies that are aware of our location down to a few millimeters and can communicate in real time with any other system we build, technologies that are constructed and operated by secretive global corporations and political superpowers, technologies that can tap into vast pools of data gathered about everyone on Earth, are frightening. Yet we are inviting them, uncritically it seems, into our lives. This, I think, is where the politics of clocks deserves our most urgent attention, although by no means to the exclusion of all the other examples we have examined in this book.

In the end, we need to look beyond the faces of our clocks to see who is inside, rather than simply obeying what these objects tell us to do. For thousands of years, people have smuggled their political agendas in clocks, or as clocks, knowing that we are more likely to accept the instructions of an object than of other humans, for reasons I do not fully understand. Who can fight back against the relentless beat of the clock? Perhaps we should try a little harder.

To be honest, I cannot see an idyllic future for humankind. What has struck me the most, in writing this book, is the consistency of its civilization-level themes across thousands of years, once you clear away the froth of our present-day experience. So I do not think we will stop using clocks in governing people, in securing power and in controlling human behavior, often in ways that cause suffering and loss. But at least if we understand where our clocks have come from—who has made them and why—we might be better equipped to criticize and, more often, resist. We can challenge; ask difficult questions; arm ourselves with knowledge. In other words, be good citizens.

The time capsule buried at Osaka in 1970 included a message, to be read in the year 6970, written by Rin Masayuki, a fourth-grade pupil at a Tokyo primary school:

> A society where everyone lives cheerfully and happily . . . this must be the same goal for me and you, I believe. We must do our best until the next age takes over. Goodbye from 5,000 years in the past.[11]

We might not be able fundamentally to reshape the structures of civilization, although it might be interesting to try. But we can each, in our own way, make changes that leave the world a little better than when we came into it. Clocks are tools at our disposal. If other people are using them to control our lives, we can use them ourselves, too, to good effect. And it is about time that we did, because, otherwise, our time might run out.

Acknowledgments

I have benefited from the generous support, wisdom and expertise of so many people in writing this book.

Jack Ramm was the commissioning editor at Viking who first invited me to consider writing a book about time and steered me deftly and with constant support through the commissioning process and into writing. Tom Killingbeck later took over the reins from Jack and has been a never-failing guide since. I am enormously grateful to both, and to their wonderful colleagues across Penguin Random House. Thanks, too, to the teams at W. W. Norton and other publishers for taking this story worldwide. And thanks to my former colleague Wendy Burford for her crucial support in the commissioning stages of the project.

I would like to thank all the scholars whose deep research and thought-provoking publications have enabled me to piece this account together. Thanks, too, to the countless collectors and other people involved in the study of timekeeping that I have met at lectures, society meetings and study visits who have so generously shared their expertise. And I am forever grateful to all the librarians, archivists, museum staff and other specialists who have given me their help and support over many years.

Jonathan Betts, James Nye, Keith Scobie-Youngs and Richard Stenning have been pillars of strength, not just on this project but many others. Their expertise is second to none and their warm support is what has made this all possible. I thank them too for sharing their own research (in particular James's unpublished work on Johann Antel) and for reading the manuscript and offering their wise opinion. It was we five who holidayed in Venice, Siena and Amsterdam, in part so that I could visit clocks and frescoes for this book. They are patient as well as wise! At The Clockworks, James Nye's electrical horology center in London, the resident conservators James Harris and Johan ten Hoeve have inspired and taught me so much over the years I have been involved there.

For kindly arranging for me to see some of the clocks presented here in the flesh, I have many people to thank (and in most cases many more were involved than those I met—thanks to them all). In Thuwal, Marie-Laure Boulot and Ana Catarina Matias. In Washington, DC, John Gray and Jaya Kaveeshwar. At Jodrell Bank, Peter Linstead-Smith and Ian Morison. In Amsterdam, Rob Memel and Henk Verhoef. At Piccadilly Circus, Mark Crangle and Keith Scobie-Youngs. At Greenwich Park, Derrick Spurr. And at Chioggia thanks go to Marisa Addomine and Daniele Pons as well as to all the people of the city that we met during a wonderful visit. I will never forget the pageant performed for us by the children and staff of the Istituto Comprensivo Chioggia 1.

The clock specialists at NPL have been a constant support to me throughout my career, and for this project I am particularly grateful to Ali Ashkhasi, Charlotte Blake, Marivon Corbel, Sam Gresham, Leon Lobo and Peter Whibberley for allowing me to learn about and see the NPL clocks providing time to the financial sector. To help me understand the client side, the former chair of the London Stock Exchange, Donald Brydon, kindly arranged for me to meet the LSE experts Phil Crossley, Denzil Jenkins, Kaushalya Kularatnam and Liam Smith, in a session organized by Lesley Bird and Nikhil Rathi. I learned a huge amount at our meeting and was grateful for the time taken to explain such a complex situation in language I could understand. Elsewhere in the time and frequency industry, I have been helped by countless experts, with Bob Cockshott, Charles Curry and John Pottle being particular stars.

Moving from fast clocks to the slowest, I must thank the people of the Long Now Foundation for everything they have done for me since we first met in 2000. Stewart Brand, Danny Hillis and Brian Eno ignited in me a new way of thinking, and particular gratitude goes to Alexander Rose, who has been an inspiration throughout. I met Laura Williams in front of the Long Now clock the same year; Aluna, and Laura's friendship, has helped sustain me ever since. And I am truly grateful to Jem Finer, whose projects—not just Longplayer, though that was how we met—have changed the way I see and experience the world.

More widely, I have benefited hugely over several years from the wisdom and generosity of David Gilbert and many other colleagues at the Department of Geography, Royal Holloway, University of London, where I hold a research associateship. I am also grateful to my former colleagues at the Science Museum Group and Royal Museums Greenwich, including trustees and advisers, who have helped shape and guide my thoughts over the past quarter century, and still do. Andrew Nahum must get a special mention here. He took a chance on me in 1997 and I remain in his debt.

Debasish Das kindly let me see his ongoing research into clock towers in India. The late Joyce Christie gave me invaluable insights into Mary Dixon, the Golden Voice competition finalist from Jarrow who helped bring her up. David Perrett generously shared unpublished research into Henry Ford's 1928 visit to London. Simon Schaffer offered vivid insights into racial discrimination in the early years of the Cape of Good Hope observatory.

I would like to thank everyone who has permitted the reproduction of copyrighted material in this book. For their particular help with the provision of images, I thank Bonhams, Charles Frodsham and Co., James Harris, the Panasonic Corporation, and the executors of the estate of

Joyce Christie: John Robinson, Paul Hutchinson and Ian Smith. I must also express my gratitude to John Agard, through Georgia Holmes at Caroline Sheldon Literary Agency, for allowing me to reproduce his poem about the Rugby clock. Meeting John at Greenwich in 2007 was a great inspiration, setting my research on a new path, though I have a long way yet to travel along it.

Finally, this book is dedicated, with love, to my family.

Notes

INTRODUCTION

1. Quoted in "Attachment C: Background Information Related to the Report of the Completion of the Fact-Finding Investigation Regarding the Shooting Down of Korean Air Lines Boeing 747 (Flight KE 007) on 31 August 1983," in *State Letter 93/68* (Montreal: International Civil Aviation Organization, 1993), 14–16.
2. "Transcript of President Reagan's Address on Downing of Korean Airliner," *The New York Times*, 6 September 1983, 15.

1. ORDER

1. Attributed to the playwright Plautus, translated and quoted in Robert Hannah, *Time in Antiquity* (London and New York: Routledge, 2009), 82.
2. The playwright Alkiphron, translated and quoted in ibid., 82.
3. Cassiodorus, translated and quoted in Paulo Forlati, "Roman Solar-Acoustic Clock in Verona," *Antiquarian Horology* 9, no. 2 (March 1975): 199.
4. Translated and quoted in ibid., 201.
5. Quoted in Wu Hung, "Monumentality of Time: Giant Clocks, the Drum Tower, the Clock Tower," in *Monuments and Memory, Made and Unmade*, ed. Robert Nelson and Margaret Olin (Chicago and London: University of Chicago Press, 2003), 114.

6. Quoted in ibid., 115.

7. Marisa Addomine, "A Fourteenth-Century Italian Turret Clock," *Antiquarian Horology* 37, no. 2 (June 2016): 222.

8. Galeazzo Gattari, quoted in Richard Goy, *Chioggia and the Villages of the Venetian Lagoon* (Cambridge: Cambridge University Press, 1985), 37.

9. "London, Wednesday, November 30, 1892," *The Times*, November 30, 1892, 9.

10. Giordano Nanni, *The Colonisation of Time: Ritual, Routine and Resistance in the British Empire* (Manchester and New York: Manchester University Press, 2012), 16.

11. Anon., "Delhi Clock Tower," *The Builder*, February 14, 1874, 130.

12. Anon., "Clock Tower, Lucknow," *The Builder*, August 1, 1885, 170.

13. Sanjay Srivastava, *Constructing Post-Colonial India: National Character and the Doon School* (London and New York: Routledge, 1998), 47–48.

14. Fani Efendi, quoted in Mehmet Bengü Uluengin, "Secularizing Anatolia Tick by Tick: Clock Towers in the Ottoman Empire and the Turkish Republic," *International Journal of Middle East Studies* 42, no. 1 (February 2010): 20.

15. Hung, "Monumentality of Time," 128.

2. FAITH

1. This and subsequent phrases quoted in Ibn al-Razzāz al-Jazarī, *The Book of Knowledge of Ingenious Mechanical Devices*, trans. Donald Hill (Dordrecht: D. Reidel, 1974), 17–41.

2. Quoted in John Beckmann, *A History of Inventions, Discoveries, and Origins*, Vol. I (London: Henry G. Bohn, 1846), 345.

3. Translated and quoted in Otto Kurz, *European Clocks and Watches in the Near East* (London: The Warburg Institute, 1975), 17.

4. Quoted in Lynn Thorndike, *The Sphere of Sacrobosco and Its Commentators* (Chicago: University of Chicago Press, 1949), 230.

5. John North, *God's Clockmaker: Richard of Wallingford and the Invention of Time* (New York: Continuum, 2005), 320.

6. Edward Davis and Michael Hunter, eds., *Robert Boyle: A Free Enquiry into the Vulgarly Received Notion of Nature* (Cambridge: Cambridge University Press, 1996 [1686]), 13.

7. Quoted in John Castle, *Remarkable Clocks* (unpublished manuscript, in Arthur Mitchell archive, Antiquarian Horological Society, London, 1951), 10.5.

8. Ibid., 10.16.

9. Quoted in Jaroslav Folta, "Clockmaking in Medieval Prague," *Antiquarian Horology* 23, no. 5 (Autumn 1997): 408.

10. Both quotations in Charles George, "A Social Interpretation of English Puritanism," *Journal of Modern History* 25, no. 4 (December 1953): 338–39.

11. Lewis Mumford, *Technics and Civilization* (Chicago: University of Chicago Press, 1934), 14.

12. Benjamin Franklin, "Advice to a Young Tradesman, Written by an Old One," in George Fisher, *The American Instructor; or, Young Man's Best Companion*, 9th edn. (Philadelphia, 1748), 375.

13. Quoted in Lauren Frayer, "Saudis Want 'Mecca Time' to Replace GMT," *AOL News*, 11 August 2010.

14. Ziauddin Sardar, "The Destruction of Mecca," *The New York Times*, September 30, 2014.

3 · VIRTUE

1. Translated and quoted in Charles Drover, "Sand-Glass 'Sand': Historical, Analytical and Practical, Part I—Historical," *Antiquarian Horology* 3, no. 3 (June 1960): 64.

2. Thomas Carlyle, *Sartor Resartus*, ed. Archibald MacMechan (Boston: Ginn & Company, 1900), 199.

3. Quoted in Quentin Skinner, "Ambrogio Lorenzetti: The Artist as Political Philosopher," *Proceedings of the British Academy* 72 (1986): 50.

4. Quoted in Lynn White, "The Iconography of *Temperantia* and the Virtuousness of Technology," in Theodore Rabb and Jerrold Seigel (eds.), *Action and Conviction in Early Modern Europe: Essays in Memory of E. H. Harbison* (Princeton, NJ: Princeton University Press, 1969), 211.

5. Quoted in ibid., 209.

6. Translated and quoted in Kristen Lippincott, *The Story of Time* (London: Merrell Holberton, 1999), 175.

7. Charles Johnson, *A General History of the Pyrates* (London, 1724), 259.

4 · MARKETS

1. Paul Hemelryk, representing Hornby, Hemelryk and Company, quoted in House of Commons, *Report and Special Report from the Select Committee on the Daylight Saving Bill, Together with the Proceedings of the Committee, Minutes of Evidence, and Appendix* (London: HMSO, 1908), 118.

2. "The New Liverpool Cotton Exchange: Description of the Building," *Liverpool Daily Post and Mercury*, July 2, 1906, 9.

3. Benjamin Franklin, "Advice to a Young Tradesman, Written by an Old One," in George Fisher, *The American Instructor; or, Young Man's Best Companion*, 9th edn. (Philadelphia, 1748), 375.

4. Quoted in Jon Cartwright, "Time Traders," *Physics World*, July 2018.

5. KNOWLEDGE

1. Joseph Tieffenthaler, 1751, quoted in Virendra Nath Sharma, *Sawai Jai Singh and His Astronomy*, 2nd edn. (Delhi: Motilal Banarsidass Publishers Private Ltd., 2015), 124.

2. Sharma, *Sawai Jai Singh*, 290–91.

3. Quoted in Susan Johnson-Roehr, "The Spatialization of Knowledge and Power at the Astronomical Observatories of Sawai Jai Singh II, c. 1721–1743 CE" (Urbana: University of Illinois at Urbana-Champaign, 2011), 228.

4. Quoted in Raymond Mercier, "Account by Joseph Dubois of Astronomical Work under Jai Singh Sawāʾī," *Indian Journal of History of Science* 28, no. 2 (1993): 162.

5. Quoted in "Full Text: 'Bin Laden's Message,'" BBC News, 12 November 2002, http://news.bbc.co.uk/1/hi/world/middle_east/2455845.stm.

6. Translated from the French original quoted in René Taton, "Les Origines et les débuts de l'Observatoire de Paris," *Vistas in Astronomy* 20 (1976): 67.

7. Bernard Lovell, *The Story of Jodrell Bank* (New York and Evanston, Ill.: Harper & Row, 1968), 196.

6. EMPIRES

1. Quoted in Brian Warner, *Astronomers at the Royal Observatory Cape of Good Hope: A History with Emphasis on the Nineteenth Century* (Cape Town: A. A. Balkema, 1979), 34.

2. Adam Smith, *An Inquiry into the Nature and Causes of the Wealth of Nations*, Vol. II (London, 1776), 235.

3. Robert Percival, *An Account of the Cape of Good Hope* (London, 1804), 9.

4. David Gill, *A History and Description of the Royal Observatory, Cape of Good Hope* (London: HMSO, 1913), x.

5. Quoted in Warner, *Astronomers at the Royal Observatory*, 34.

6. Quoted in Brian Warner, *Royal Observatory, Cape of Good Hope 1820–1831: The Founding of a Colonial Observatory* (Dordrecht: Kluwer Academic Publishers, 1995), 180–81.

7. John Cannon, "Masons," *Cape Town Gazette and African Advertiser*, August 12, 1825, 2.

8. Quoted in Warner, *Royal Observatory, Cape of Good Hope 1820–1831*, 182.

9. Quoted in Brian Warner, *Charles Piazzi Smyth, Astronomer-Artist: His Cape Years 1835–1845* (Cape Town: A. A. Balkema, 1981), 16.

10. Quoted in ibid., 72.

11. Walter Raleigh, *The Works of Sir Walter Ralegh*, Vol. VIII: *Miscellaneous Works* (Oxford: Oxford University Press, 1829), 325.

12. *A Collection of All Queen Anne's Speeches, Messages &c from Her Accession to the Throne, to Her Demise* (London, 1714), 52.

13. *The History and Proceedings of the House of Commons of England, Tome V* (London, 1742), 144.

14. Edwin Dunkin, "The Royal Observatory, Greenwich. A Day at the Observatory (Part 2)," *The Leisure Hour*, January 16, 1862, 40.

15. "The New Eros," *The Graphic*, December 15, 1928, 452.

16. William Mitchell, *Time and Weather by Wireless* (London: The Wireless Press, 1923), vii.

17. David Gill, "Report of the Proceedings of Cape of Good Hope, Royal Observatory," *Monthly Notices of the Royal Astronomical Society* 64 (February 1904): 304.

7. MANUFACTURE

1. "Extraordinary Escape from Death at Bennett's Clock Warehouse, Cheapside," *The Sun*, May 27, 1865, 3.

2. "London Gossip," *Stirling Observer*, September 7, 1865, 4.

3. Sir John Bennett, Ltd., *Gog and Magog: The House of Bennett 1750–Present Day* (London: Sir John Bennett, 1920), 8.

4. "Our Gallery: Sir John Bennett, FRAS," *Watchmaker, Jeweller, and Silversmith* 1, no. 8 (August 5, 1875): 57.

5. Quoted in Richard Harvey, "Bennett, Sir John (1814–1897)," *Oxford Dictionary of National Biography* (online edn.), September 2004.

6. "The 'City Observatory,'" *City Press*, June 17, 1865, 5.

7. "Death of Sir John Bennett: Departure of a Quaint Figure," *Pall Mall Gazette*, July 6, 1897, 7.

8. John Bennett, "To the Editor of the City Press," *City Press*, 9 September 1865, 6.

9. Quoted in Albert Musson and Eric Robinson, "The Origins of Engineering in Lancashire," *Journal of Economic History* 20, no. 2 (June 1960): 218.

10. John Kennedy, "Observations on the Rise and Progress of the Cotton Trade in Great Britain, Particularly in Lancashire and the Adjoining Counties (Read Before the Literary and Philosophical Society of Manchester, November 3, 1815)," in *Miscellaneous Papers on Subjects Connected with the Manufactures of Lancashire* (Manchester, 1849), 13.

11. Quoted in Musson and Robinson, "The Origins of Engineering in Lancashire," 221.

12. British Horological Institute, "Discussion Meetings," *Horological Journal*, February 1, 1860, 80.

13. Mr. Walters, quoted in Alan Midleton, "The History of the BHI (Part 1): Clerkenwell and the Angry 1850s," *Horological Journal*, July 2007, 271.

14. Henry Ward, "Making Watches by Machinery," Hansard, House of Commons, Debated March 31, 1843, vol. 68, cols. 273–85.

15. Anon., "The Misplaced Horological Institute," *Horological Journal*, October 1889, 31.

16. "Our Gallery: Sir John Bennett, FRAS," 56.

17. Quoted in Joseph Wickham Roe, *English and American Tool Builders* (New Haven, Conn.: Yale University Press, 1916), 215.

18. "Jottings," *Horological Journal*, August 1897, 163–64.

19. Quoted in "Gog and Magog," *Hull Daily Mail*, October 21, 1929, 9.

20. "The History of a Lost Trade: An Interview with Sir John Bennett," *Pall Mall Gazette*, December 13, 1886, 2.

21. David Landes, *Revolution in Time: Clocks and the Making of the Modern World*, rev. edn. (London: Viking, 2000), 318.

22. Ibid., 308.

23. Ford Motor Company (advertisement), "Needles and Nails Made His First Watch Tools . . . ," *Boys' Life*, February 1945, 15.

24. Henry Ford, *My Life and Work* (Garden City, NY: Doubleday, Page and Company, 1922), 24.

25. Adam Ferguson, *An Essay on the History of Civil Society*, 2nd edn. (London, 1768), 280.

26. Andrew Nahum, caption for Plate 8, in Peter J. T. Morris, *Science for the Nation: Perspectives on the History of the Science Museum* (Basingstoke: Palgrave Macmillan, 2010).

8. MORALITY

1. Anthony Dingle, *The Campaign for Prohibition in Victorian England* (London: Croom Helm, 1980), 8.

2. John Kennaway, quoted in "Intoxicating Liquors Bill," Hansard, House of Commons, Debated June 5, 1874, vol. 219, col. 1090.

3. Sidney Webb, "Preface (1902)," in B. L. Hutchins and A. Harrison, *A History of Factory Legislation*, new edn. (London: P. S. King & Son, 1907), v.

4. *An Act to Amend the Laws Relating to Labour in Factories*, June 6, 1844, section XXVI, 7 Victoria Ch. 15.

5. *Factory and Workshop Act*, 1901, 1 Edward VII Ch. 22, p. 21.

6. BT Archives: POST 30/531, *Synchronisation etc. of Clocks by Electric Current*, letter, Secretary, Oldham Master Cotton Spinners' Association to District Manager of Telephones, GPO, SE Lancs District, Rochdale, December 4, 1913.

7. Gertrude Magrane, "Recollections of William Willett," in *Petts Wood 21st Birthday Festival Week Souvenir Programme, 16–22 May 1948*, 1948.

8. Quoted in House of Commons, *Report and Special Report from the Select Committee on the Daylight Saving Bill, Together with the Proceedings of the Committee, Minutes of Evidence, and Appendix*, 116.

9. Quoted in ibid., 49.

10. *The Times*, May 4, 1911, 7.

11. Ibid., May 22, 1916, 9.

12. Quoted in House of Commons, *Report and Special . . . on the Daylight Saving Bill*, 76.

9. RESISTANCE

1. *Dalkeith Advertiser*, May 22, 1913.

2. See, for instance, *Westminster Gazette*, May 21, 1913.

3. George Woodcock, "The Tyranny of the Clock," *War Commentary for Anarchism* 5, no. 10 (March 1944).

4. E. P. Thompson, "Time, Work-Discipline, and Industrial Capitalism," *Past & Present* 38, no. 1 (1 December 1967): 95.

5. Quoted in ibid., 86.

6. Quoted in Maurice Thomas, *The Early Factory Legislation* (Leigh-on-Sea, UK: Thames Bank Publishing Co., 1948), 39.

7. Benjamin Hargreaves, quoted in James Haslam, *Accrington and Church Industrial Co-Operative Society Ltd: History of Fifty Years' Progress* (Manchester: Co-Operative Newspaper Society, 1910), 200.

8. William Baron, quoted in Bob Haye, "Struck by Several Sods: Violence and the 1878 Blackburn Weavers' Strike" (BA Dissertation), http://www.Cottontown.Org/The%20Cotton%20Industry/Cotton%20Industry%2018th%20to%2020th%20Century/Pages/Cotton-Riots-1878.Aspx, n.d.

9. Mark M. Smith, *Mastered by the Clock: Time, Slavery, and Freedom in the American South* (Chapel Hill: University of North Carolina Press, 1997), 95.

10. Quoted in ibid., 121.

11. Michael Bakunin, *God and the State* (London: Freedom Press, 1910), 41.

12. Ibid., 18.

13. Ibid., 20.

14. Cambridge University Library, Royal Greenwich Observatory Archives: RGO 7/58, *Papers on Greenwich Park*, letter, Astronomer Royal to HM Office of Works, January 27, 1885.

15. Ibid., letter, Astronomer Royal to Admiralty, January 5, 1894.

16. Quoted in *Illustrated London News*, February 17, 1894, 195.

17. *The Times*, February 20, 1894, 5.

18. Ibid., February 27, 1894, 8.

19. Quoted in John Merriman, *The Dynamite Club* (London: JR Books, 2009), 187.

10. IDENTITY

1. BT Archives: "Speaking Clock" subject file, letter, John Masefield to Kingsley Wood, June 5, 1935.

2. Ibid.: " 'Voice of Gold' Competition" score sheet.

3. Quoted in "The Girl with the Golden Voice," *Post Office Magazine*, August 1935, 263.

4. Quoted in *Croydon Advertiser and Surrey County Reporter*, June 29, 1935.

5. BT Archives: "Speaking Clock" subject file, advance summary of speech to be given by the Postmaster General, July 24, 1936.

6. *Croydon Advertiser and Surrey County Reporter*, June 29, 1935; *Evening News*, June 21, 1935[?]; *Croydon Times and Surrey County Mail*, June 29, 1935.

7. BT Archives: POST 33/4799: Public Relations Department circular, March 27, 1935.

8. Ibid.: letter, London Telephone Service to Public Relations Department, June 13, 1935 (my italics).

9. *The Times*, May 25, 1935, 9.

10. BT Archives: "Speaking Clock" subject file, advance summary of speech to be given by the GPO Staff Engineer, July 1936.

11. "The Girl with the Golden Voice," 263.

12. British Pathé, *Time Please* (film, 1938).

13. V&A Theatre Collections, scrapbook, "Prince of Wales 1935."

14. GPO Film Unit, *At the 3rd Stroke* (film, 1939).

15. Lewis Carroll, *Alice's Adventures in Wonderland* (New York: D. Appleton and Co., 1866), 101–2.

16. Reproduced in Henry Hawken, *Trumpets of Glory: Fourth of July Orations, 1786–1861* (Granby, Conn.: Salmon Brook Historical Society, 1976), 259.

17. Felix Meier, quoted in Michael O'Malley and Carlene Stephens, "Clockwork History: Monumental Clocks and the Depiction of the American Past," *NAWCC Bulletin*, February 1990, 8.

18. O'Malley and Stephens, "Clockwork History," 8.

19. Quoted in the *Guardian*, August 7, 2015.

20. Quoted in the *Telegraph*, May 5, 2018.

21. Douglas Schoen and Michael Rowan, *The Threat Closer to Home: Hugo Chávez and the War against America* (New York: Free Press, 2009), 44.

22. Quoted in Jonathan Hassid and Bartholomew Watson, "State of Mind: Power, Time Zones and Symbolic State Centralization," *Time & Society* 23, no. 2 (2014): 180.

23. Quoted in ibid., 181.

24. Quoted in ibid., 182.

25. European Union (Withdrawal Agreement) Bill, Hansard, House of Commons, Debated January 9, 2020, vol. 669, col. 711.

26. *Daily Express*, January 16, 2020.

27. Ian Dunt, "Week in Review: The Madness over Big Ben's Bongs Is a Symbol of the Horror to Come," January 17, 2020, www.politics.co.uk/blogs/2020/01/17/week-in-review-the-madness-over-big-ben-s-bongs-is-a-symbol.

11. WAR

1. "Cosmic Confrontation," *Measure: For the Men and Women of Hewlett-Packard*, March 1972, 5.

2. Stephen Powers and Brad Parkinson, "The Origins of GPS: Article Published in May & June 2010 Issues of GPS World," 2010, 17, https://www.u-blox.com/sites/default/files/the_origins_of_gps.pdf.

3. Andreas Krieg and Jean-Marc Rickli, "Surrogate Warfare: The Art of War in the 21st Century?," *Defence Studies* 18, no. 2 (2018): 115.

4. Derek Gregory, "The Everywhere War," *Geographical Journal* 177, no. 3 (September 2011): 238.

5. Office of Science and Technology Policy, National Security Council, *Fact Sheet: US Global Positioning System Policy* (Washington, DC: 1996).

6. Richard Schwartz, quoted in Jack Loughran, "Interview: The Creators of GPS," *E&T Magazine*, February 15, 2019.

7. Chuck Horner, quoted in Greg Milner, *Pinpoint: How GPS Is Changing Our World* (New York: W. W. Norton & Company, 2016), 47.

8. Quoted in Jim Quinn, "I Had to Sell This to the Air Force, Because the Air Force Didn't Want It," *Invention and Technology* 20, no. 2 (Fall 2004).

9. Quoted in Milner, *Pinpoint*, 58.

10. Chris Whitty and Mark Walport, *Satellite-Derived Time and Position: A Study of Critical Dependencies* (London: Government Office for Science, 2018), 29.

11. John Garamendi and Dana Goward, quoted in Paul Tullis, "The World Economy Runs on GPS. It Needs a Backup Plan," *Bloomberg Businessweek*, July 25, 2018, https://www.bloomberg.com/news/features/2018-07-25/the-world-economy-runs-on-gps-it-needs-a-backup-plan.

12. PEACE

1. Matsushita Electric Industrial Co. Ltd., *The Official Record of Time Capsule Expo '70: A Gift to the People of the Future from the People of the Present Day* (Osaka: Matsushita Electric Industrial Co. Ltd., 1980), 229.

2. Ibid., 3.

3. Ibid., 231.

4. Quoted in Stewart Brand, *The Clock of the Long Now: Time and Responsibility* (London: Phoenix, 2000), 2.

5. Ibid., 2.

6. Quoted in ibid., 4–5.

7. Brian Eno, "Bells and Their History, and The Long Now Foundation" (CD insert), in *January 07003: Bell Studies for The Clock of The Long Now* (CD), 2003, 4.

8. Kevin Kelly, "Clock in the Mountain," accessed July 7, 2020, longnow.org/clock; Jonas Salk, "Are We Being Good Ancestors?," *World Affairs: The Journal of International Issues* 1, no. 2 (December 1992): 16–18.

9. Brand, *Clock of the Long Now*, 53.

10. Alexander Rose, "How to Build Something That Lasts 10,000 Years," BBC Future, June 11, 2019, https://www.bbc.com/future/article/20190611-how-to-build -something-that-lasts-10000-years.

11. Matsushita Electric Industrial Co. Ltd., *Official Record of Time Capsule Expo '70*, 232.

Selected Sources

INTRODUCTION

Easton, Richard, and Eric Frazier. *GPS Declassified: From Smart Bombs to Smartphones.* Lincoln, Nebr.: Potomac Books, 2013.

Gordon, Michael. "Ex-Soviet Pilot Still Insists KAL 007 Was Spying." *The New York Times,* December 9, 1996.

Hersh, Seymour. *The Target Is Destroyed: What Really Happened to Flight 007 and What America Knew about It.* London: Faber and Faber, 1986.

Powers, Stephen, and Brad Parkinson. "The Origins of GPS: Article Published in May & June 2010 Issues of GPS World," 2010. https://www.u-blox.com/sites/default/files/the_origins_of_gps.pdf.

Proceedings: First International Symposium on Precise Positioning with the Global Positioning System (Rockville, Maryland, April 15–19, 1985), Vol. I. Washington, DC: US Department of Commerce, 1985.

Proceedings of the Sixteenth Annual Precise Time and Time Interval (PTTI) Applications and Planning Meeting (a Meeting Held at the NASA Goddard Space Flight Center, Greenbelt, Maryland, November 27–29, 1984). NASA, 1984.

State Letter 93/68: Destruction of Korean Air Lines Flight 007 on 31 August 1983. Montreal: International Civil Aviation Organization, 1993.

Stephens, Carlene. "Time in Place: Cold War Clocks in the American West," in *Where*

Minds and Matters Meet: Technology in California and the West, ed. Volker Janssen, 321–45. Berkeley: University of California Press, 2012.

"Transcript of President Reagan's Address on Downing of Korean Airliner." *The New York Times*, September 6, 1983.

1. ORDER

Addomine, Marisa. "A Fourteenth-Century Italian Turret Clock." *Antiquarian Horology* 37, no. 2 (June 2016): 213–22.

Baillie, G. H., H. Alan Lloyd and F. A. B. Ward. *The Planetarium of Giovanni de Dondi*. London: Antiquarian Horological Society, 1974.

Bedini, Silvio. *The Trail of Time: Time Measurement with Incense in East Asia*. Cambridge: Cambridge University Press, 1994.

Cache, Bernard. "The Tower of the Winds of Andronikos of Kyrros: An Inaugural and Surprisingly Contemporary Building." *Architectural Theory Review* 14, no. 1 (2009): 3–18.

Das, Debasish. *India's Pre-Independence Era Public Clocks* (unpublished manuscript), 2020.

Davison, Graeme. *The Unforgiving Minute: How Australia Learned to Tell the Time*. Melbourne: Oxford University Press, 1993.

Ekinci, Ekrem Buğra. "Ottoman-Era Clock Towers Telling Time from Balkans to Middle East." *Daily Sabah*, January 6, 2017.

Forlati, Paulo. "Roman Solar-Acoustic Clock in Verona." *Antiquarian Horology* 9, no. 2 (March 1975): 198–201.

Goy, Richard. *Chioggia and the Villages of the Venetian Lagoon*. Cambridge: Cambridge University Press, 1985.

Gratwick, A. S. "Sundials, Parasites, and Girls from Boeotia." *The Classical Quarterly* 29, no. 2 (1979): 308–23.

Hannah, Robert. *Time in Antiquity*. London and New York: Routledge, 2009.

Hoff, M. "The Early History of the Roman Agora at Athens." *Bulletin Supplement (University of London, Institute of Classical Studies)*, no. 55, *The Greek Renaissance in the Roman Empire: Papers from the Tenth British Museum Classical Colloquium*, 1989, 1–8.

Hung, Wu. "Monumentality of Time: Giant Clocks, the Drum Tower, the Clock Tower," in *Monuments and Memory, Made and Unmade*, ed. Robert Nelson and Margaret Olin, 107–32. Chicago and London: University of Chicago Press, 2003.

Kontogiannis, Nikos. "Review of Pamela Webb, *The Tower of the Winds in Athens: Greeks, Romans, Christians and Muslims: Two Millennia of Continual Use*." *Speculum* 94, no. 2 (April 2019): 600–602.

Lusnia, Susann. "Battle Imagery and Politics on the Severan Arch in the Roman Forum," in *Representations of War in Ancient Rome,* ed. Sheila Dillon and Katherine Welch, 272–99. Cambridge: Cambridge University Press, 2006.

Metcalf, Thomas. "Architecture and the Representation of Empire: India, 1860–1910." *Representations* 6 (Spring 1984): 37–65.

Nanni, Giordano. *The Colonisation of Time: Ritual, Routine and Resistance in the British Empire.* Manchester and New York: Manchester University Press, 2012.

Noble, Joseph, and Derek de Solla Price. "The Water Clock in the Tower of the Winds." *American Journal of Archaeology* 72, no. 4 (October 1968): 345–55.

Papalexandrou, Nassos. "Review of Pamela Webb, *The Tower of the Winds in Athens: Greeks, Romans, Christians, and Muslims: Two Millennia of Continual Use.*" *Bryn Mawr Classical Review,* 2019.04.31 (2019).

Pattenden, Philip. "A Late Sundial at Aphrodisias." *The Journal of Hellenic Studies* 101 (1981): 101–12.

Pliny. *The Natural History,* Vol. II, trans. John Bostock and H. T. Riley. London: Henry G. Bohn, 1855.

Porter, Andrew. *The Oxford History of the British Empire,* Vol. II: *The Nineteenth Century.* Oxford and New York: Oxford University Press, 1999.

Ravagnan, Sergio. *La Guida: Chioggia, Sottomarina, Isolaverde e Dintorni.* Chioggia: Città di Chioggia, 2016.

Robinson, Henry. "The Tower of the Winds and the Roman Market-Place." *American Journal of Archaeology* 47, no. 3 (1943): 291–305.

Rosenhain, Margaret. "A Land of Lots of Time." *Walkabout,* January 1972, 55–57.

Singer, Sean. *Clock Towers, Blended Modernity, and the Emergence of Ottoman Time* (MA Dissertation). Bloomington: Indiana University, 2013.

Srivastava, Sanjay. *Constructing Post-Colonial India: National Character and the Doon School.* London and New York: Routledge, 1998.

Turner, A. J. "Water-Clocks," in *The Time Museum: Catalogue of the Collection,* Vol. I: *Time Measuring Instruments,* Part 3: *Water-Clocks, Sand-Glasses, Fire-Clocks,* 1–44. Rockford, Ill.: The Time Museum, 1984.

Uluengin, Mehmet Bengü. "Secularizing Anatolia Tick by Tick: Clock Towers in the Ottoman Empire and the Turkish Republic." *International Journal of Middle East Studies* 42, no. 1 (February 2010): 17–36.

Webb, Pamela. "Review of Hermann Kienast, *Der Turm der Winde in Athen. Archäologische Forschungen,*" Vol. 30. *Bryn Mawr Classical Review,* 2015.09.35 (2015).

———. *The Tower of the Winds in Athens: Greeks, Romans, Christians, and Muslims: Two Millennia of Continual Use.* Philadelphia: American Philosophical Society Press, 2017.

Wolf, Caroline "Olivia" M. "Marking Time, Marking Movement: Mexico City's Ottoman Clock Tower as a Transnational Expression of Immigrant Identity." *Hemisphere: Visual Cultures of the Americas* 9, no. 1 (2016): 24–43.

2. FAITH

Addomine, Marisa. "Italian Astronomical Clocks as Public Astrological Machines," in *Heaven and Earth United: Instruments in Astrological Contexts*, ed. Richard Dunn, Silke Ackermann and Giorgio Strano, 133–52. Leiden and Boston: Brill, 2018.

Beckmann, John. *A History of Inventions, Discoveries, and Origins*, Vol. I. London: Henry G. Bohn, 1846.

Bedini, Silvio. "The Scent of Time: A Study of the Use of Fire and Incense for Time Measurement in Oriental Countries." *Transactions of the American Philosophical Society* 53, no. 5 (1963): 1–51.

———. *The Trail of Time: Time Measurement with Incense in East Asia*. Cambridge: Cambridge University Press, 1994.

Bromley, John. *The Clockmakers' Library: The Catalogue of the Books and Manuscripts in the Library of the Worshipful Company of Clockmakers*. London: Sotheby Parke Bernet Publications, 1977.

Carey, Hilary. *Courting Disaster: Astrology at the English Court and University in the Later Middle Ages*. New York: Palgrave Macmillan, 1992.

Castle, John. *Remarkable Clocks* (unpublished manuscript, in Arthur Mitchell archive, Antiquarian Horological Society, London), 1951.

Cipolla, Carlo. *Clocks and Culture 1300–1700*. London: Collins, 1967.

Davis, Edward, and Michael Hunter, eds. *Robert Boyle: A Free Enquiry into the Vulgarly Received Notion of Nature*. Cambridge: Cambridge University Press, 1996 [1686].

Desborough, Jane. *The Changing Face of Early Modern Time, 1550–1770*. Cham: Palgrave Macmillan, 2019.

Dohrn-van Rossum, Gerhard. *History of the Hour: Clocks and Modern Temporal Orders*. Chicago: University of Chicago Press, 1996.

Fischer, Karl. "The Ancient Town Hall Clock of Prague in Changing Times." *Antiquarian Horology* 4, no. 6 (March 1964), 174–75.

Folta, Jaroslav. "Clockmaking in Medieval Prague." *Antiquarian Horology* 23, no. 5 (Autumn 1997): 405–17.

Franklin, Benjamin. "Advice to a Young Tradesman, Written by an Old One," in George Fisher, *The American Instructor; or, Young Man's Best Companion*, 375–78, 9th edn. Philadelphia, 1748.

George, Charles. "A Social Interpretation of English Puritanism." *Journal of Modern History* 25, no. 4 (December 1953): 327–42.

Jagger, Cedric. *Royal Clocks: The British Monarchy and Its Timekeepers 1300–1900*. London: Robert Hale, 1983.

———. *The World's Great Clocks and Watches*. London: Hamlyn Publishing Group, 1977.

Jazarī, Ibn al-Razzāz al-. *The Book of Knowledge of Ingenious Mechanical Devices*, trans. Donald Hill. Dordrecht: D. Reidel, 1974.

King, David. "Astronomy in the Service of Islam," in *Handbook of Archaeoastronomy and Ethnoastronomy*, ed. Clive Ruggles, 181–96. New York: Springer-Verlag, 2015.

Kurz, Otto. *European Clocks and Watches in the Near East*. London: The Warburg Institute, 1975.

Landes, David. *Revolution in Time: Clocks and the Making of the Modern World*. Cambridge, Mass.: Harvard University Press, 1983.

Lloyd, H. Alan. *Some Outstanding Clocks over Seven Hundred Years 1250–1950*. London: Leonard Hill Books, 1958.

McCluskey, Stephen. "Astronomy in the Service of Christianity," in *Handbook of Archaeoastronomy and Ethnoastronomy*, ed. Clive Ruggles, 165–79. New York: Springer-Verlag, 2015.

Morrill, John. "Cromwell, Oliver (1599–1658)." *Oxford Dictionary of National Biography* (online edn.), September 2015.

Mumford, Lewis. *Technics and Civilization*. Chicago: University of Chicago Press, 1934.

North, John. *God's Clockmaker: Richard of Wallingford and the Invention of Time*. New York: Continuum, 2005.

Oestmann, Günther. "The Strasbourg Cathedral Clock." *Antiquarian Horology* 25, no. 1 (September 1999), 50–63.

Pacey, Arnold. *The Maze of Ingenuity: Ideas and Idealism in the Development of Technology*. London: Allen Lane, 1974.

Pleasure, Myron. "Time in Jewish Tradition." *Antiquarian Horology* 8, no. 6 (March 1974), 619–24.

Thompson, David. *Clocks*. London: British Museum Press, 2004.

———. *Watches*. London: British Museum Press, 2008.

Thorndike, Lynn. *The Sphere of Sacrobosco and Its Commentators*. Chicago: University of Chicago Press, 1949.

Turner, A. J. "Donald Hill and Arabic Water-Clocks." *Antiquarian Horology* 27, no. 2 (December 2002), 206–13.

———. "Fire-Clocks," in *The Time Museum: Catalogue of the Collection*, Vol. I: *Time Measuring Instruments*, Part 3: *Water-Clocks, Sand-Glasses, Fire-Clocks*, 117–22. Rockford, Ill.: The Time Museum, 1984.

———. "Water-Clocks," in *The Time Museum: Catalogue of the Collection*, Vol. I: *Time Measuring Instruments*, Part 3: *Water-Clocks, Sand-Glasses, Fire-Clocks*, 1–44. Rockford, Ill.: The Time Museum, 1984.

Wainwright, Oliver. "Mecca's Mega Architecture Casts Shadow over Hajj." *Guardian*, October 23, 2012.

White, Lynn. *Medieval Technology and Social Change.* Oxford: Oxford University Press, 1962.

Whitestone, Sebastian. "Time before the Oscillator: Horology in the Thirteenth-Century Manuscripts of the Libros Del Saber." *Antiquarian Horology* 40, no. 4 (December 2019): 487–500.

Whitrow, Gerald James. "Time and Timing: The Astronomical and Historical Development." *Die Naturwissenschaften* 64, no. 3 (1977): 105–12.

———. *Time in History: Views of Time from Prehistory to the Present Day.* Oxford and New York: Oxford University Press, 1988.

3. VIRTUE

Aked, C. K., and J. R. A. Aked. "Sand Glasses." *Horological Journal*, October 1977, 3–10.

Ascheri, Mario, and Bradley Franco. *A History of Siena: From Its Origins to the Present Day.* Abingdon: Routledge, 2020.

Balmer, R. T. "The Operation of Sand Clocks and Their Medieval Development." *Technology and Culture* 19, no. 4 (October 1978): 615–32.

Boucheron, Patrick. *The Power of Images: Siena, 1338.* Cambridge: Polity Press, 2018

Boullin, David, and Anne Nairne-Clark. "Hourglasses on Some English and Scottish Tombstones." *Antiquarian Horology* 22, no. 3 (Autumn 1995): 252–53.

Cohen, Kathleen. *Metamorphosis of a Death Symbol: The Transi Tomb in the Late Middle Ages and the Renaissance.* Berkeley and Los Angeles: University of California Press, 1973.

Cohen, Simona. "The Early Renaissance Personification of Time and Changing Concepts of Temporality." *Renaissance Studies* 14, no. 3 (September 2000): 301–28.

Coole, P. G. "Sand-Glass 'Sand': Historical, Analytical and Practical, Part III—Practical." *Antiquarian Horology* 3, no. 3 (June 1960): 72.

Dohrn-van Rossum, Gerhard. *History of the Hour: Clocks and Modern Temporal Orders.* Chicago: University of Chicago Press, 1996.

Drover, Charles. "Sand-Glass 'Sand': Historical, Analytical and Practical, Part I—Historical." *Antiquarian Horology* 3, no. 3 (June 1960): 62–67.

Howat, John, and Trudie Roberts. "The Hourglass and Trade Symbols on Scottish Tombstones." *Antiquarian Horology* 22, no. 6 (Summer 1996): 544–45.

Lippincott, Kristen. *The Story of Time.* London: Merrell Holberton, 1999.

Macey, Samuel. "The Changing Iconography of Father Time," in *The Study of Time*, Vol. III, ed. Julius Fraser, Nathaniel Lawrence and David Park, 540–77. New York: Springer-Verlag, 1978.

McCormick, Finbar. "The Symbols of Death and the Tomb of John Forster in Tydavnet, Co. Monaghan." *Clogher Record* 11, no. 2 (1983): 273–86.

Michel, Henry. "Some New Documents in the History of Horology." *Antiquarian Horology* 3, no. 10 (March 1962): 288–91.

Panofsky, Erwin. *Studies in Iconology: Humanistic Themes in the Art of the Renaissance.* New York: Oxford University Press, 1939.

Pizan, Christine de. *Othea's Letter to Hector,* ed. Renate Blumenfeld-Kosinski and Earl Jeffrey Richards. Toronto: Iter Press, 2017.

Ricasoli, Corinna. " 'Memento Mori' in Baroque Rome." *Studies: An Irish Quarterly Review* 104, no. 416 (Winter 2015/2016): 456–67.

Sabine, P. A. "Sand-Glass 'Sand': Historical, Analytical and Practical, Part II—Analytical." *Antiquarian Horology* 3, no. 3 (June 1960): 68–72.

Skinner, Quentin. "Ambrogio Lorenzetti: The Artist as Political Philosopher." *Proceedings of the British Academy* 72 (1986): 1–56.

———. "Ambrogio Lorenzetti's Buon Governo Frescoes: Two Old Questions, Two New Answers." *Journal of the Warburg and Courtauld Institutes* 62 (1999): 1–28.

Steiner, George. "The Uncommon Reader [1978]," in *No Passion Spent: Essays 1978–1995.* London: Faber and Faber, 1996.

Thon, Peter. "Bruegel's *The Triumph of Death* Reconsidered." *Renaissance Quarterly* 21, no. 3 (Autumn 1968): 289–99.

Tomalin, Marcus. " 'An Invaluable Acquisition': Sandglasses in Romantic Literature." *European Romantic Review* 28, no. 6 (2017): 729–49.

Turner, A. J. " 'The Accomplishment of Many Years': Three Notes towards a History of the Sand-Glass." *Annals of Science* 39, no. 2 (1982): 161–72.

———. "Sand-Glasses," in *The Time Museum: Catalogue of the Collection,* Vol. I: *Time Measuring Instruments,* Part 3: *Water-Clocks, Sand-Glasses, Fire-Clocks,* 75–113. Rockford, Ill.: The Time Museum, 1984.

White, Lynn. "The Iconography of *Temperantia* and the Virtuousness of Technology," in Theodore Rabb and Jerrold Seigel, eds., *Action and Conviction in Early Modern Europe: Essays in Memory of E. H. Harbison,* 197–219. Princeton, NJ: Princeton University Press, 1969.

4. MARKETS

Abrahamse, J. E. *De grote uitleg van Amsterdam: Stadsontwikkeling in de zeventiende eeuw* (PhD Thesis). Amsterdam: Amsterdam Institute for Humanities Research, 2010.

Appleby, Joyce. *The Relentless Revolution.* New York and London: W. W. Norton & Company, 2010.

Aquilina, Matteo, and Carla Ysusi. *Are High-Frequency Traders Anticipating the Order Flow? Cross-Venue Evidence from the UK Market* (Occasional Paper 16). London: Financial Conduct Authority, 2016.

Aron, Jacob. "Atomic Time to Rule High-Speed Trading." *New Scientist*, no. 1186 (April 19, 2014).

Braudel, Fernand. *Civilization and Capitalism, 15th–18th Century*, Vol. II: *The Wheels of Commerce*. London: William Collins Sons & Co., 1982.

Cartwright, Jon. "Time Traders." *Physics World*, July 2018.

Dohrn-van Rossum, Gerhard. *History of the Hour: Clocks and Modern Temporal Orders*. Chicago: University of Chicago Press, 1996.

Groot, A. H. de. *The Ottoman Empire and the Dutch Republic: A History of the Earliest Diplomatic Relations 1610–1630*. Rev. edn. Leiden: Nederlands Instituut voor het Nabije Oosten, 2012.

Hall, Nigel. "The Liverpool Cotton Market: Britain's First Futures Market." *Transactions of the Historic Society of Lancashire and Cheshire* 149 (1999): 99–117.

Houghton, Alistair. "Liverpool's Lost Gem: Why Stunning Cotton Exchange Facade Was Flattened." *Liverpool Echo*, November 30, 2016.

Kocka, Jürgen. *Capitalism: A Short History*. Princeton, NJ, and Oxford: Princeton University Press, 2016.

Lewis, Michael. *Flash Boys: Cracking the Money Code*. London: Allen Lane, 2014.

NPL. *A Complete Guide to Time Stamping Regulations in the Financial Sector*. Teddington: NPL, 2018.

———. *Time Traceability for the Finance Sector: Fact Sheet*. Teddington: NPL, 2018.

Petram, Lodewijk. *The World's First Stock Exchange*. New York and Chichester: Columbia University Press, 2014.

Smith, M. F. J. *Tijd-affaires in effecten aan de Amsterdamsche beurs*. 'S-Gravenhage: Martinus Nijhoff, 1919.

Thornbury, Walter. "The Royal Exchange," in *Old and New London: A Narrative of Its History, Its People, and Its Places*, Vol. I, 494–513. London, Paris and New York: Cassell, Petter, & Galpin, 1878.

———. "The Stock Exchange," in *Old and New London: A Narrative of Its History, Its People, and Its Places*, Vol. I, 473–94. London, Paris and New York: Cassell, Petter, & Galpin, 1878.

Wikipedia (Netherlands). "Beurs van Hendrick de Keyser," July 25, 2019.

———. "Oosterkerk (Amsterdam)," August 18, 2019.

5. KNOWLEDGE

Cocroft, Wayne, and Roger Thomas. *Cold War: Building for Nuclear Confrontation 1946–1989*, ed. P. S. Barnwell. Swindon: Historic England, 2003.

Connor, Elizabeth. "The Cassini Family and the Paris Observatory." *Astronomical Society of the Pacific Leaflets* 5, no. 218 (1947): 146–53.

Digby, Simon. "Review of V. S. Bhatnagar, *Life and Times of Sawai Jai Singh, 1688–1743*."

Bulletin of the School of Oriental and African Studies, University of London 39, no. 1 (1976): 193–95.

Fazlıoğlu, İhsan. "The Samarqand Mathematical-Astronomical School: A Basis for Ottoman Philosophy and Science." *Journal for the History of Arabic Science* 14 (2008): 3–68.

Forbes, Eric. "The European Astronomical Tradition: Its Transmission into India, and Its Reception by Sawai Jai Singh II." *Indian Journal of History of Science* 17, no. 2 (1982): 234–43.

Frazier, Ian. "Invaders: Destroying Baghdad." *New Yorker*, April 25, 2005.

Grahn, Sven. "Jodrell Bank's Role in Early Space Tracking Activities." Jodrell Bank Centre for Astrophysics, September 2008. www.jb.man.ac.uk/history/tracking/part1 .html and www.jb.man.ac.uk/history/tracking/part2.html.

Hung, Wu. "Monumentality of Time: Giant Clocks, the Drum Tower, the Clock Tower," in *Monuments and Memory, Made and Unmade*, ed. Robert Nelson and Margaret Olin, 107–32. Chicago and London: University of Chicago Press, 2003.

Johnson-Roehr, Susan. "Observatories of Sawai Jai Singh II," in Clive Ruggles, ed., *Handbook of Archaeoastronomy and Ethnoastronomy*, 2017–28. New York: Springer-Verlag, 2015.

———. "The Spatialization of Knowledge and Power at the Astronomical Observatories of Sawai Jai Singh II, c. 1721–1743 CE." Champaign: University of Illinois at Urbana-Champaign, 2011.

Lovell, Bernard. *The Story of Jodrell Bank*. New York and Evanston, Ill.: Harper & Row, 1968.

MacDougall, Bonnie G. "Jantar Mantar: Architecture, Astronomy, and Solar Kingship in Princely India." *The Cornell Journal of Architecture* 5 (1996): 16–33.

Mercier, Raymond. "Account by Joseph Dubois of Astronomical Work under Jai Singh Sawā'ī." *Indian Journal of History of Science* 28, no. 2 (1993): 157–66.

Needham, Joseph, Wang Ling and Derek de Solla Price. *Heavenly Clockwork: The Great Astronomical Clocks of Medieval China*. 2nd edn. Cambridge: Cambridge University Press, 1986.

Nomination of the Jantar Mantar, Jaipur, for Inclusion on World Heritage List. Paris: UNESCO, 2010.

Pettigrew, William. *Freedom's Debt: The Royal African Company and the Politics of the Atlantic Slave Trade, 1672–1752*. Chapel Hill: University of North Carolina Press, 2013.

Ruggles, Clive, and Michel Cotte. *Heritage Sites of Astronomy and Archaeoastronomy in the Context of the UNESCO World Heritage Convention: A Thematic Study* (printed edn.). Paris: International Council on Monuments and Sites/International Astronomical Union, 2011.

Scott, W. R. "The Constitution and Finance of the Royal African Company of England from Its Foundation till 1720." *American Historical Review* 8, no. 2 (January 1902): 241–59.

Sharma, Virendra Nath. "Jai Singh, His European Astronomers and the Copernican Revolution." *Indian Journal of History of Science* 18, no. 1 (1983): 333–44.

———. *Sawai Jai Singh and His Astronomy.* 2nd edn. Delhi: Motilal Banarsidass Publishers Private Ltd., 2015.

Solla Price, Derek de. "Astronomy's Past Preserved at Jaipur." *Natural History: The Journal of the American Museum of Natural History* 73, no. 6 (July 1964): 48–53.

Taton, René. "Les Origines et les débuts de l'Observatoire de Paris." *Vistas in Astronomy* 20 (1976): 65–71.

Tezcan, Baki. "Some Thoughts on the Politics of Early Modern Ottoman Science." *Osmanlı Araştırmaları: The Journal of Ottoman Studies* 36 (2010): 135–56.

Xiaochun, Sun, and Han Yi. "The Northern Song State's Financial Support for Astronomy." *East Asian Science, Technology, and Medicine* 38 (2013–14): 17–53.

6. EMPIRES

Andrewes, William, ed. *The Quest for Longitude.* Cambridge, Mass.: Collection of Historical Scientific Instruments, Harvard University, 1996.

Astronomical Society of Southern Africa: Historical Section. "Royal Observatory at the Cape of Good Hope," n.d. http://assa.saao.ac.za/sections/history/observatories/royal_cape_obs/.

Betts, Jonathan. *Marine Chronometers at Greenwich.* Oxford: Oxford University Press, 2017.

Bisset, W. M. "Cape Town's Time-Guns." *Scientia Militaria, South African Journal of Military Studies* 14, no. 4 (1984): 67–71.

Brück, Hermann, and Mary Brück. *The Peripatetic Astronomer: The Life of Charles Piazzi Smyth.* Bristol and Philadelphia: Adam Hilger, 1988.

Dunkin, Edwin. "The Royal Observatory, Greenwich. A Day at the Observatory (Part 1)." *The Leisure Hour*, January 9, 1862.

———. "The Royal Observatory, Greenwich. A Day at the Observatory (Part 2)." *The Leisure Hour*, January 16, 1862.

Dunn, Richard, and Rebekah Higgitt, eds. *Navigational Enterprises in Europe and Its Empires, 1730–1850.* Basingstoke: Palgrave Macmillan, 2015.

Eglash, Ron. "Broken Metaphor: The Master–Slave Analogy in Technical Literature." *Technology and Culture* 48, no. 2 (April 2007): 360–69.

Ellis, William. "Lecture on the Treatment of Chronometers, at the Royal Observatory, Greenwich." *Horological Journal*, April 1, 1866, 85–92.

Gill, David. *A History and Description of the Royal Observatory, Cape of Good Hope.* London: HMSO, 1913.

Heller, Michael. "'Outposts of Britain': The General Post Office and the Birth of a Corporate Iconic Brand, 1930–1939." *European Journal of Marketing* 50, no. 3/4 (2016): 358–76.

Homes, Caitlin. "The Astronomer Royal, the Hydrographer and the Time Ball: Collaborations in Time Signalling 1850–1910." *British Journal for the History of Science* 42, no. 3 (September 2009): 381–406.

Howse, Derek. *Greenwich Time and the Longitude*. London: Philip Wilson, 1997.

Laing, J. D., ed. *The Royal Observatory at the Cape of Good Hope 1820–1970: A Sesquicentennial Offering*. Cape Town: Royal Observatory, Cape of Good Hope, 1970.

Lords Commissioners of the Admiralty. *List of Time Signals, Established in Various Parts of the World*. London: Hydrographic Department, Admiralty, 1908.

Mackenzie, Theodore. "Historic Determinations of the Longitude of the Cape (Part 1)." *Monthly Notices of the Astronomical Society of South Africa* 11, no. 4 (April 1952): 34–35.

Maunder, Walter. *The Royal Observatory Greenwich: A Glance at Its History and Work*. London: Religious Tract Society, 1900.

May, W. E. "How the Chronometer Went to Sea." *Antiquarian Horology* 9, no. 6 (March 1976): 638–63.

Nockolds, Susanna. "Early Timekeepers at Sea (Part 1)." *Antiquarian Horology* 4, no. 4 (September 1963): 110–13.

———. "Early Timekeepers at Sea (Part 2)." *Antiquarian Horology* 4, no. 5 (December 1963): 148–52.

Percival, Robert. *An Account of the Cape of Good Hope*. London: C. and R. Baldwin, 1804.

Piazzi Smyth, Charles. "Present State of the Longitude Question in Navigation," in *Two Lectures Delivered before the Edinburgh Chamber of Commerce*. Edinburgh, 1859.

"The Planet-Watchers of Greenwich." *Harper's New Monthly Magazine* 1, no. 2 (July 1850): 233–37.

Rooney, David. *Ruth Belville: The Greenwich Time Lady*. London: National Maritime Museum, 2008.

Ruggles, Clive, and Michel Cotte. *Heritage Sites of Astronomy and Archaeoastronomy in the Context of the UNESCO World Heritage Convention: A Thematic Study* (printed edn.). Paris: International Council on Monuments and Sites/International Astronomical Union, 2011.

Schaffer, Simon. "Smyth's Elevation: Victorian Astronomy's Vertical Empire (Keynote Lecture)," at *Stars, Pyramids and Photographs: Charles Piazzi Smyth, 1819–1900*, Royal Society of Edinburgh, September 4, 2019.

Seemann, Ute. "Dutch and British Coastal Fortifications at the Cape of Good Hope (1665 to 1829)." South African History Online, April 2010. https://www.sahistory.org.za/place/dutch-and-british-coastal-fortifications-cape-good-hope-1665-1829.

"The Time-Ball of St Helena." *Nautical Magazine* 4 (1835): 658–60.

Warner, Brian. *Astronomers at the Royal Observatory Cape of Good Hope: A History with Emphasis on the Nineteenth Century*. Cape Town: A. A. Balkema, 1979.

———. *Charles Piazzi Smyth, Astronomer-Artist: His Cape Years 1835–1845*. Cape Town: A. A. Balkema, 1981.

———. *Royal Observatory, Cape of Good Hope 1820–1831: The Founding of a Colonial Observatory*. Dordrecht: Kluwer Academic Publishers, 1995.

Wauchope, Robert. "Time Signals for Chronometers." *Nautical Magazine* 5 (1836): 460–64.

7. MANUFACTURE

Baines, Edward. *History of the Cotton Manufacture in Great Britain*. London: H. Fisher, R. Fisher, and P. Jackson, 1835.

Bennetts of Greenwich. "Bennetts" History," n.d. http://www.bennettsofgreenwich.co .uk/bennetts-history/.

Bryan, Ford. *Henry's Attic: Some Fascinating Gifts to Henry Ford and His Museum*. Detroit: Wayne State University Press, 2006.

Bud, Robert, Simon Niziol, Timothy Boon and Andrew Nahum. *Inventing the Modern World: Technology Since 1750*. London: Dorling Kindersley, 2000.

Church, Roy. "Nineteenth-Century Clock Technology in Britain, the United States, and Switzerland." *Economic History Review* 28, no. 4 (November 1975): 616–30.

Davies, Alun C. "Horology at International Industrial Exhibitions, 1851–1900." *Antiquarian Horology* 33, no. 5 (September 2012), 591–608.

———. "The Ingold Episode Revisited: English Watchmaking's Pyrrhic Victory." *Antiquarian Horology* 31, no. 5 (September 2009): 637–54.

———. "An Invasion in Time: American Horology and the British Market." *Antiquarian Horology* 30, no. 6 (June 2008), 829–44.

Ferguson, Adam. *An Essay on the History of Civil Society*. 2nd edn. London, 1768.

Ford, Henry. *My Life and Work*. Garden City, NY: Doubleday, Page and Company, 1922.

Good, Richard. "An Enamelled Gold Minute Repeating Striking Watch with Arms, Crest and Cypher of Sir John Bennett." *Horological Journal*, January 1984, 12–15 and 22.

Harvey, Richard. "Bennett, Sir John (1814–1897)." *Oxford Dictionary of National Biography* (online edn.), September 2004.

Hoopes, Penrose. *Early Clockmaking in Connecticut*. New Haven: Tercentenary Commission of the State of Connecticut, 1934.

Hounshell, David. *From the American System to Mass Production, 1800–1932*. Baltimore: Johns Hopkins University Press, 1984.

Kennedy, John. "Observations on the Rise and Progress of the Cotton Trade in Great Britain, Particularly in Lancashire and the Adjoining Counties (Read before the Literary and Philosophical Society of Manchester, November 3, 1815)," in *Miscellaneous Papers on Subjects Connected with the Manufactures of Lancashire*, 5–25. Manchester, 1849.

Lamont-Brown, Raymond. *John Brown: Queen Victoria's Highland Servant*. Stroud: History Press, 2000.

Landes, David. *Revolution in Time: Clocks and the Making of the Modern World*. Rev. edn. London: Viking, 2000.

"Making Watches by Machinery." Hansard, House of Commons, Debated on March 31, 1843, Vol. 68, cols. 273–85.

Midleton, Alan. "The History of the BHI (Part 1): Clerkenwell and the Angry 1850s." *Horological Journal*, July 2007, 268–71.

———. "The History of the BHI (Part 2): The Early Years." *Horological Journal*, August 2007, 312–15.

Muir, Diana. *Reflections in Bullough's Pond: Economy and Ecosystem in New England*. Hanover, NH, and London: University Press of New England, 2000.

Murphy, John Joseph. "Entrepreneurship in the Establishment of the American Clock Industry." *Journal of Economic History* 26, no. 2 (June 1966): 169–86.

Musson, Albert, and Eric Robinson. "The Origins of Engineering in Lancashire." *Journal of Economic History* 20, no. 2 (June 1960): 209–33.

Palmer, Brooks. *The Book of American Clocks*. New York: Macmillan Publishing Co., 1950.

Perrett, David. "Henry Ford's 1928 English Holiday, Part 1—In Search of Newcomen Engines." *International Journal for the History of Engineering & Technology* 88, no. 1 (2018): 37–56.

Roe, Joseph Wickham. *English and American Tool Builders*. New Haven, Conn.: Yale University Press, 1916.

"The Rumbustious Days When the Institute Was Formed—and Some Outstanding BHI Personalities." *Horological Journal*, September 1958, 556–68.

Sharpe, Henry. *A Measure of Perfection: The History of Brown and Sharpe*. Excerpts from an Address to the Newcomen Society of England, May 4, 1949.

Sir John Bennett, Ltd. *Gog and Magog: The House of Bennett 1750–Present Day*. London: Sir John Bennett, 1920.

Stowers, A. "The Preservation of Historic Machinery and Its Problems." *Transactions of the Newcomen Society* 28, no. 1 (1951): 207–23.

8. MORALITY

Bartky, Ian. *One Time Fits All: The Campaigns for Global Uniformity*. Stanford: Stanford University Press, 2007.

Downing, Michael. *Spring Forward: The Annual Madness of Daylight Saving*. Washington, DC: Shoemaker & Hoard, 2005.

Galison, Peter. *Einstein's Clocks, Poincaré's Maps: Empires of Time*. London: Hodder & Stoughton, 2003.

Howse, Derek. *Greenwich Time and the Longitude*. London: Philip Wilson, 1997.

Morus, Iwan Rhys. *Nikola Tesla and the Electrical Future*. London: Icon Books, 2019.

Nye, James. *Johann Antel 1866–1930: Pioneering Electric Horology in Early 20th Century Brno* (unpublished manuscript), 2005.

Nye, James, and David Rooney. "'Such Great Inventors as the Late Mr Lund': An Introduction to the Standard Time Company, 1870–1970." *Antiquarian Horology* 30, no. 4 (December 2007): 501–23.

Prerau, David. *Saving the Daylight: Why We Put the Clocks Forward*. London: Granta Books, 2005.

Rooney, David. *Ruth Belville: The Greenwich Time Lady*. London: National Maritime Museum, 2008.

———. "Speculative Building, Planner-Preservationists and Daylight: Commemorations of William Willett, 1918–1948." *Construction Information Quarterly* 11, no. 4 (2009): 207–11.

Rooney, David, and James Nye. "'Greenwich Observatory Time for the Public Benefit': Standard Time and Victorian Networks of Regulation." *British Journal for the History of Science* 42, no. 1 (March 2009): 5–30.

Sedlak, Jan. "History MT." National Theatre Brno, 22 December 2008. https://web.archive.org/web/20081222025225/http://www.ndbrno.cz/en/about-us/theatre-buildings/mahen-theatre/history-of-mahen-theatre/history-mt/.

9. RESISTANCE

Archer, John E. *Social Unrest and Popular Protest in England, 1780–1840*. Cambridge: Cambridge University Press, 2000.

Atkins, Keletso E. "'Kafir Time': Preindustrial Temporal Concepts and Labour Discipline in Nineteenth-Century Colonial Natal." *Journal of African History* 29, no. 2 (1988): 229–44.

Bakunin, Michael. *God and the State*. London: Freedom Press, 1910.

Barrows, Adam. "'The Shortcomings of Timetables': Greenwich, Modernism, and the Limits of Modernity." *MFS Modern Fiction Studies* 56, no. 2 (Summer 2010): 262–89.

Bartky, Ian. *One Time Fits All: The Campaigns for Global Uniformity*. Stanford: Stanford University Press, 2007.

Brück, Hermann. *The Story of Astronomy in Edinburgh: From Its Beginnings until 1975*. Edinburgh: Edinburgh University Press, 1983.

Conroy, Mark. "The Panoptical City: The Structure of Suspicion in *The Secret Agent*." *Conradiana* 15, no. 3 (1983): 203–17.

Gantt, Jonathan. "Irish-American Terrorism and Anglo-American Relations, 1881–1885." *Journal of the Gilded Age and Progressive Era* 5, no. 4 (October 2006): 325–57.

Harling, Nick. "The Cotton Riots of 1878." http://www.Cottontown.Org/The%20 Cotton%20Industry/Cotton%20Industry%2018th%20to%2020th%20Century/ Pages/Cotton-Riots-1878.Aspx, n.d.

Haslam, James. *Accrington and Church Industrial Co-Operative Society Ltd: History of Fifty Years' Progress.* Manchester: Co-Operative Newspaper Society, 1910.

Haye, Bob. "Struck by Several Sods: Violence and the 1878 Blackburn Weavers' Strike" (BA Dissertation). http://www.Cottontown.Org/The%20Cotton%20Industry/Cotton%20 Industry%2018th%20to%2020th%20Century/Pages/Cotton-Riots-1878.Aspx, n.d.

Horn, Jeff. "Machine-Breaking in England and France during the Age of Revolution." *Labour/Le Travail* 55 (Spring 2005): 143–66.

Krishnan, Shekhar. *Empire's Metropolis: Money, Time & Space in Colonial Bombay, 1870–1930* (PhD Thesis). Cambridge, Mass.: Massachusetts Institute of Technology, 2013.

Merriman, John. *The Dynamite Club.* London: JR Books, 2009.

Piper, Karen. *Cartographic Fictions.* New Brunswick, NJ: Rutgers University Press, 2002.

Pollard, Sidney. "Factory Discipline in the Industrial Revolution." *The Economic History Review* 16, no. 2 (1963): 254–71.

Randall, Adrian J. *Before the Luddites: Custom, Community, and Machinery in the English Woollen Industry, 1776–1809.* Cambridge: Cambridge University Press, 1991.

Riddell, Fern. "The Weaker Sex? Violence and the Suffragette Movement." *History Today* 65, no. 3 (March 2015).

Royal Observatory Edinburgh Archives: Contemporary Correspondence and Press Cuttings Relating to 1913 Bombing.

Salamé, Richard. *Clocks and Empire: An Indian Case Study.* Providence, RI: Brown University, 2014. https://library.brown.edu/ugresearchprize/Salame-ClocksandEmpire.pdf.

Smith, Mark M. *Mastered by the Clock: Time, Slavery, and Freedom in the American South.* Chapel Hill: University of North Carolina Press, 1997.

Thompson, E. P. "Time, Work-Discipline, and Industrial Capitalism." *Past & Present* 38, no. 1 (December 1, 1967): 56–97.

Woodcock, George. "The Tyranny of the Clock." *War Commentary for Anarchism* 5, no. 10 (March 1944).

10. IDENTITY

Anthony, Scott. *Public Relations and the Making of Modern Britain: Stephen Tallents and the Birth of a Progressive Media Profession.* Manchester: Manchester University Press, 2012.

Anthony, Scott, and James Mansell, eds. *The Projection of Britain: A History of the GPO Film Unit.* Basingstoke: Palgrave Macmillan, 2011.

Bennett, Oliver. *Daylight Saving Bill 2010–11: Research Paper 10/78*. London: House of Commons Library, 2010.

De Carle, Donald. *British Time*. London: C. Lockwood, 1947.

Hassid, Jonathan, and Bartholomew Watson. "State of Mind: Power, Time Zones and Symbolic State Centralization." *Time & Society* 23, no. 2 (2014): 167–94.

Lowenthal, David. *The Past Is a Foreign Country—Revisited*. Cambridge: Cambridge University Press, 2015.

McCrossen, Alexis. "'Conventions of Simultaneity': Time Standards, Public Clocks, and Nationalism in American Cities and Towns, 1871–1905." *Journal of Urban History* 33, no. 2 (January 2007): 217–53.

McKay, Chris. *Big Ben: The Great Clock and the Bells at the Palace of Westminster*. Oxford: Oxford University Press, 2010.

O'Malley, Michael, and Carlene Stephens. "Clockwork History: Monumental Clocks and the Depiction of the American Past." *NAWCC Bulletin*, February 1990, 3–15.

Rooney, David. *Ruth Belville: The Greenwich Time Lady*. London: National Maritime Museum, 2008.

Schoen, Douglas, and Michael Rowan. *The Threat Closer to Home: Hugo Chávez and the War against America*. New York: Free Press, 2009.

Stevens, Tim. "Governing the Time of the World," in *Time, Temporality and Global Politics*, ed. Andrew Hom, Christopher McIntosh, Alasdair McKay and Liam Stockdale, 59–72. Bristol: E-International Relations Publishing, 2016.

11. WAR

Allnutt, Jason, Dhananjay Anand, Douglas Arnold, et al. *Timing Challenges in the Smart Grid* (NIST Special Publication 1500-08). Gaithersburg: National Institute of Standards and Technology, 2017.

Curry, Charles. *Sentinel Project: Report on GNSS Vulnerabilities*. Lydbrook: Chronos Technology Ltd., 2014.

Dudziak, Mary. *War Time: An Idea, Its History, Its Consequences*. Oxford: Oxford University Press, 2012.

Easton, Richard. "Who Invented the Global Positioning System?" *The Space Review*, May 2006.

Easton, Richard, and Eric Frazier. *GPS Declassified: From Smart Bombs to Smartphones*. Lincoln, Nebr.: Potomac Books, 2013.

Easton, Roger L. "Global Navigation Flies High." *Physics World*, October 2007, 34–38.

Engineering and Technology History Wiki. "Brad Parkinson: An Oral History Conducted in 1999 by Michael Geselowitz, IEEE History Center, Hoboken, NJ," 1999. https://ethw.org/Oral-History:Brad_Parkinson.

Gregory, Derek. "The Everywhere War." *Geographical Journal* 177, no. 3 (September 2011): 238–50.

Humud, Carla, Christopher Blanchard and Mary Beth Nikitin. *Armed Conflict in Syria: Overview and U.S. Response*. Washington, DC: Congressional Research Service, 2019.

Jones, Tony. *Splitting the Second: The Story of Atomic Time*. Bristol: Institute of Physics, 2000.

Krieg, Andreas, and Jean-Marc Rickli. "Surrogate Warfare: The Art of War in the 21st Century?" *Defence Studies* 18, no. 2 (2018): 113–30.

Loughran, Jack. "Interview: The Creators of GPS." *E&T Magazine*, February 15, 2019.

Milner, Greg. *Pinpoint: How GPS Is Changing Our World*. New York: W. W. Norton & Company, 2016.

Office of Science and Technology Policy, National Security Council. *Fact Sheet: US Global Positioning System Policy*, Washington, DC, 1996.

Pace, Scott, Gerald Frost, Irving Lachow, et al. *The Global Positioning System: Assessing National Policies*. Santa Monica: RAND, 1995.

Powers, Stephen, and Brad Parkinson. "The Origins of GPS: Article Published in May & June 2010 Issues of GPS World," 2010. https://www.u-blox.com/sites/default/files/the_origins_of_gps.pdf.

Proceedings of the Fourth Annual NASA and Department of Defense Precise Time and Time Interval (PTTI) Planning Meeting. NASA, 1972.

Proceedings of the Sixteenth Annual Precise Time and Time Interval (PTTI) Applications and Planning Meeting (a Meeting Held at the NASA Goddard Space Flight Center, Greenbelt, Maryland, November 27–29, 1984), NASA, 1984.

Quinn, Jim. "I Had to Sell This to the Air Force, Because the Air Force Didn't Want It." *Invention and Technology* 20, no. 2 (Fall 2004).

Riley, William J. "A History of the Rubidium Frequency Standard." IEEE UFFC-S History. http://ieee-uffc.org/aboutus/history/a-history-of-the-rubidium-frequency-standard.pdf, December 2019, 1–35.

Stephens, Carlene. "Time in Place: Cold War Clocks in the American West," in *Where Minds and Matters Meet: Technology in California and the West*, ed. Volker Janssen, 321–45. Berkeley: University of California Press, 2012.

Sturdevant, Rick. "NAVSTAR, the Global Positioning System: A Sampling of Its Military, Civil, and Commercial Impact," in *Societal Impact of Spaceflight*, ed. Stephen J. Dick and Roger D. Launius, 331–51. Washington, DC: NASA, 2007.

Teunussen, Peter, and Oliver Montenbruck, eds. *Springer Handbook of Global Navigation Satellite Systems*. Cham: Springer, 2017.

Whitty, Chris, and Mark Walport. *Satellite-Derived Time and Position: A Study of Critical Dependencies*. London: Government Office for Science, 2018.

12. PEACE

Brand, Stewart. *The Clock of the Long Now: Time and Responsibility*. London: Phoenix, 2000.

Brown, Austin. "Alexander Rose Visits Ise Shrine Reconstruction Ceremony." Blog of the Long Now, October 3, 2013. https://blog.longnow.org/02013/10/03/alexander -rose-visits-ise-shrine-reconstruction-ceremony/.

Edahiro, Junko. "Rebuilding Every 20 Years Renders Sanctuaries Eternal—the Sengu Ceremony at Jingu Shrine in Ise." *Japan for Sustainability*, August 2013. https://www .japanfs.org/en/news/archives/news_id034293.html.

Eno, Brian. *January 07003: Bell Studies for The Clock of the Long Now* (CD), 2003.

Finer, Jem. *Longplayer*. London: Artangel, 2003.

Jarvis, William. *Time Capsules: A Cultural History*. Jefferson, NC, and London: McFarland & Company, 2003.

Kelly, Kevin. "Clock in the Mountain." Accessed July 7, 2020. longnow.org/clock.

Matsushita Electric Industrial Co. Ltd. *The Official Record of Time Capsule Expo '70: A Gift to the People of the Future from the People of the Present Day*. Osaka: Matsushita Electric Industrial Co. Ltd., 1980.

Rooney, David. "Prototype 'Clock of the Long Now' (1999)," in Peter Morris, James Hart and Lesley Henderson, eds., *Milestones of Science and Technology: Making the Modern World*, 238–39. Chicago and London: KWS Publishers, 2013.

Rose, Alexander. "How to Build Something That Lasts 10,000 Years." *BBC Future*, June 11, 2019. https://www.bbc.com/future/article/20190611-how-to-build-something-that -lasts-10000-years.

———. "Long-Term Building in Japan." Blog of the Long Now, September 11, 2019. https://blog.longnow.org/02019/09/11/long-term-building-in-japan/.

Soul of Japan: An Introduction to Shinto and Ise Jingu. Tokyo: Public Affairs Headquarters for Shikinen-Sengu, 2013.

Credits

Index

Note: Page numbers in *italics* signify illustrations.